D0465170

66913

RB 155 .G3585 1992

Gene mapping

MAY 0 6 1997	DATE DUE		
DEC 16 2001			

GENE
MAPPING

GENE MAPPING

Using Law and Ethics as Guides

EDITED BY

George J. Annas
Sherman Elias

RED ROCKS
COMMUNITY COLLEGE LIBRARY

New York   Oxford
OXFORD UNIVERSITY PRESS
1992

39 95

#25050562

66913
RB
~~11~~
1 55
G3585
1992

Oxford University Press

Oxford   New York   Toronto
Delhi   Bombay   Calcutta   Madras   Karachi
Kuala Lumpur   Singapore   Hong Kong   Tokyo
Nairobi   Dar es Salaam   Cape Town
Melbourne   Auckland

and associated companies in
Berlin   Ibadan

Copyright © 1992 by Oxford University Press, Inc.

Published by Oxford University Press, Inc.,
200 Madison Avenue, New York, New York 10016

Oxford is a registered trademark of Oxford University Press

All rights reserved. No part of this publication may be reproduced,
stored in a retrieval system, or transmitted, in any form or by any means,
electronic, mechanical, photocopying, recording, or otherwise,
without the prior permission of Oxford University Press.

Library of Congress Cataloging-in-Publication Data
Gene mapping : using law and ethics as guides / edited by
George J. Annas, Sherman Elias.
p. cm.   Includes bibliographical references and index.
ISBN 0-19-507303-7
1. Human gene mapping—Moral and ethical aspects.
2. Human gene mapping—Law and Legislation.
3. Human Genome Project.
I. Annas, George J.   II. Elias, Sherman.
   [DNLM: 1. Human Genome Project.
2. Chromosome Mapping—methods.
3. Ethics, Medical.
4. Genetic Screening—United States—legislation.
5. Genetic Screening—United States—legislation.
6. Public Policy—United States.
QH 445.2 G326]
RB155.G3585   1992   174′ .25—dc20   DNLM/DLC
for Library of Congress
91-46283

9 8 7 6 5 4 3 2 1

Printed in the United States of America
on acid-free paper

To Leonard H. Glantz, J.D., friend, colleague, critic, and often silent coauthor; and to Norman Scotch, Ph.D., who has given me unqualified support throughout my career.

G. J. A.

To Joe Leigh Simpson, M.D., my teacher, colleague, and friend; and to Albert B. Gerbie, M.D., who is like a father to me.

S. E.

RED ROCKS
COMMUNITY COLLEGE LIBRARY

Preface

The 500th anniversary of the voyage of Columbus to the new world is a fitting year in which to assess the "voyage" of the Human Genome Project. The genius of Columbus was to link the old world with the new, thereby irrevocably changing both and ultimately leading to their fusion. But the encounter of the old and new worlds also led to widespread death among American natives, slavery, and epidemics. Columbus took no lawyers, philosophers, or priests with him on his initial voyage, and as the expedition's leader, his word was law. The title of our book, *Gene Mapping: Using Law and Ethics as Guides,* recounts the genome's own map metaphor and adds a simile. Of course law and ethics will not actually guide the mapping of the human genome—scientists will do this. The simile is, however, apt in the sense that unlike Columbus, the gene mappers have brought the historians, lawyers, ethicists, and philosophers with them on their journey. Their job is not to draw the map, but to speculate about its meaning, and to guide policy makers in choosing the directions that should be traveled in our genetic future.

In 1988 we became convinced that law and ethics would add little to the Human Genome Project as a public enterprise unless the legal and ethical issues involved in the project were carefully defined and prioritized early. We accordingly proposed to assemble a group of the nation's leading experts in genetics, medicine, history of science, medical law, philosophy of science, and medical ethics for an intensive workshop designed to outline where we now stand and what is needed to help get the legal and ethical studies portion of the human genome project focused and off to a reasonable start. The Center for Human Genome Research of the National Institutes of Health agreed, and the result was the first extramural workshop on this subject funded by the center. It was co-chaired by the editors, and held in Bethesda in January 1991.

The Human Genome Project has received widespread attention and acclaim. Concentration has been on the potential benefits of the project, with almost no discussion of any possible drawbacks. This view is naive and dangerous. The Human Genome Project merits much of the praise it has garnered, but it does not deserve the benefit of the doubt usually accorded it, given the tragic history of the misuse of genetic information in the United States and abroad. The social dangers inherent in the misuse of genetic information are real, and genetic discrimination is likely to be the major result of more precise genetic information. Genetic privacy will be widely threatened in employment and insurance. The way we see disease, normalcy and humanness will likely undergo significant change. Clinical medicine will increasingly become genetics based. We are more than our genes, but mapping them will cause us as a society to move into "genetics territory," and act as if our genes determine our destiny. It will take carefully considered and crafted laws and broadly supported ethical policy to prevent the misuse of human genetics from leading to the real abuse of human beings.

All this was recognized by the participants at the workshop. Rather than simply produce the proceedings of the event, we asked the participants to rework their background papers into book chapters with the aim of going beyond the workshop's initial goal (to develop a prioritized agenda for law and ethics for the Human Genome Project) by preparing a state-of-the-art paper on their topic. The goal was to produce a book that would be useful to all those involved and interested in the social-policy aspects of the Human Genome Project. Based on the thinking of a wide spectrum of experts, the design is to enable readers to gain a solid grasp of where we are now, where we should be going, and how to get there.

The chapters of the book, read together, create their own map of the current contours of social policy in the area of human genetics. The book is divided into six parts. The first part is a discussion of the scientific and social-policy aspects of the Human Genome Project. The second part addresses specific social policy implications of the Project that have not been dealt with in detail elsewhere, including the relevance of the recombinant DNA history, the eugenics legacy, military applications, and issues of race and class in the context of genetic discrimination. The third part focuses on broader philosophical issues raised by the Human Genome Project, including reductionism and determinism, the concept of disease, and what could be meant by using germline gene therapy to "improve" human beings. The fourth part is directed most specifically at the clinical implications of the new genetics, including privacy and confidentiality, genetic screening and counseling, and using cystic fibrosis as a case study to examine how the standard of care is set in medical practice. The fifth part focuses on the legal and ethical frontiers in genetics, especially in the areas of procreative liberty, patent issues, and regulatory and oversight mechanisms. The final part is devoted to a prioritized social policy research agenda for the Human Genome Project.

Columbus was driven by fifteenth-century goals that continue to pervade our era: acquisition of knowledge (sometimes called simply "progress") and the acquisition of wealth. Both these forces propel the Human Genome Project. Like Columbus, we have little real appreciation of the consequences of our future discoveries. This book does not purport to be a guidebook to the new genetics. The remaining challenge is to confront these issues openly, honestly, and with full public participation. The theme of the fifteenth century was that "man is the measure of all things." Our challenge is to profit from the fruits of the Human Genome Project while avoiding a twenty-first century in which "genes are the measure of all things."

Many people helped to make this book a reality. We want especially to thank James Watson and Eric Jeungst for both supporting this project throughout its development and writing the foreword. All the contributors participated in the January 1991 workshop in Bethesda. Other participants who helped shape this volume by their comments and contributions at the workshop were Eric Lander, Leonard Glantz, Michael Grodin, and Nancy Wexler. Bettye Weibel deserves special mention for her work in planning and managing the workshop, and Mary Lou McClellan for her work in preparing the manuscript. Finally, we are grateful to Jeffrey House at Oxford University Press who lent his support to this project from the beginning.

Boston, Massachusetts   G. J. A.
Memphis, Tennessee   S. E.
April 1992

Contents

Foreword

Doing Science in the Real World: The Role of Ethics, Law, and the Social Sciences in the Human Genome Project

James D. Watson and Eric T. Juengst

It is a twentieth-century truism that science is not done in a vacuum and should not be pursued as if it could be. Good science affects its social context, and the practical effects of good basic science are often the most wide-ranging of all. Science, in turn, is constantly affected by professional norms, social policies, and public perceptions that frame it. Doing science in the real world means anticipating those interactions and planning accordingly. By pursuing the study of the ethical, legal, and social implications of its scientific initiatives, the National Center for Human Genome Research (NCHGR) assumes its responsibility to help make that planning timely, well informed, and productive for one important piece of modern science: the Human Genome Project.

"The Human Genome Project" is shorthand for an international collection of scientific efforts to characterize the form and content of the human genome. This research will yield information—high-resolution genetic linkage and physical maps of all the human chromosomes as well as human DNA sequence data—that will be a resource for studies of gene structure and function. As such, the project is very basic science indeed: it prepares the ground for inquiries into our many remaining questions about human molecular biology.

By the same token, however, the potential for the social impact of the Human Genome Project is proportionately broad. The research it facilitates will eventually provide insights into the treatment and prevention of both inherited disorders and diseases caused by our (genetically influenced) physiological reactions to pathogens, toxins, and mutagens of external origin. It will shed light on our evolution as a species and our development as individuals. And it will serve as a starting point for researchers exploring biological elements in human behavior. Doing the Genome Project in the real world means thinking about these outcomes from the start, so that science and society can pull together to optimize the benefits of this new knowledge for human welfare and opportunity.

As the authors in this volume illustrate, the applications of genome research will raise policy questions at many levels within society. Moreover, many of the questions will become urgent long before the genome project itself is completed. For example, as Ray White and C. Thomas Caskey illustrate in Chapter 10, among the most immediate and recurrent consequences of genome research will be the development of new diagnostic and predictive tests from the localization and identification of specific disease-related genes. With the development of index and reference marker maps of the genome over the first five years of the project, the pace of gene localization and subsequent test development will accelerate. But a gene that can immediately serve as the basis for a test is only the first step in unraveling the pathogenesis of a disorder. Consequently, predictive genetic testing will typically be available well in advance of corresponding therapeutic or curative breakthroughs. The implications of acquiring and using this genetic knowledge about individuals pose a number of choices for public and professional deliberation:

- Choices for individuals and families about whether to participate in testing, with whom to share the results, and how to act on them;
- Choices for health professionals about when to recommend testing, how to ensure its quality, how to interpret the results, and to whom to disclose information;
- Choices for employers, insurers, the courts, and other social institutions about the relative value of genetic information to the kinds of decisions they must make about individuals;
- Choices for governments about how to regulate the production and use of genetic tests and the information they provide, and how to provide access to testing and counseling services;
- Choices for society about how to improve the understanding of science and its social implications at every level and increase the participation of the public in science policymaking.

These choices are not unique products of genome research, of course. They reflect issues relevant to progress in many biomedical fields. The emergence of these choices in the context of human genetics, however, does give them a special importance, which stems from several sources. First, as Evelyne Shuster (Chapter 6) and Arthur L. Caplan (Chapter 7) correctly argue, genetic explanations of health problems still carry stigmatizing connotations for many people: genetic problems are assumed to be immutable, inheritable "taints" that intrinsically implicate the bearer's identity. Such overly deterministic interpretations of genetic concepts inappropriately raise the stakes of testing and can eliminate the self-determination that testing should facilitate for individuals. In the hands of others, these interpretations can lead to stigmatization and discrimination as well. This problem is exacerbated by the fact that, until our understanding of genetic disease improves, much of the health planning that genetic testing facilitates concerns reproduction. And as John A. Robertson notes in Chapter 13, clinical geneticists find themselves working with individuals and families on very private decisions, and policy makers must face choices with potentially contentious political implications. Moreover, our choices today on these matters are made against the backdrop of past experiences. As both Patricia A. King (Chapter 5) and Robert N. Proctor (Chapter 4) properly remind us, both the relatively recent problems of sickle cell screening in the 1970s and the more distant tragedy of the American eugenics movement in the 1930s continue to underscore the importance of moving carefully in this area.

For choices by families, professions, institutions, and governments to be sound in the face of the challenges this complicated context imposes, people at every level need to be as well informed as possible. Part of the information decision makers need is scientific, and it is the responsibility of the genomics community to provide that information. Indeed, by clarifying the nature and limits of genetic causation and by providing clinical handles on genetic disease processes, scientific research using the Genome Project's tools may go far to mitigate the concerns about these choices. But sound decision making in this area needs to be well informed in other ways as well: informed by the experiences of families at risk for genetic disease, by accurate accounts of public perceptions and historical precedents, by well-researched and articulated policy alternatives, by religious perspectives and ethical considerations. The purpose of the NCHGR Program on Ethical, Legal, and Social Implications (ELSI) is to provide support for the work involved in uncovering these forms of social information and in bringing it to the people who can use it.

The ELSI program broadens the disciplinary and professional range of those involved in the Human Genome Project well beyond the biological research community. Because of this commitment, the Center supports research in the humanities and social sciences side by side with molecular biology and is advised by clinicians and patient advocates as well as basic scientists. One dramatic example of the results is the Center's decision, as a result of its ELSI program activities, to take the lead in organizing the support of the National Institutes of Health (NIH) for careful clinical assessments of testing and counseling for cystic fibrosis carrier status. These pilot studies will provide precedent-setting information about many of the issues discussed in this volume as noted by the editors in their essay with Joe Leigh Simpson (Chapter 11). Without the wider vision the ELSI program affords, however, these studies might never have gained a place on the Human Genome Project's research agenda.

The American public, through the NCHGR, is investing substantially in the Human Genome Project's efforts to anticipate its social consequences. The ELSI program provided $1.6 million in support for this work in fiscal year (FY) 1990 and anticipates doubling that amount in FY 1991. For FY 1992, the Center's budget plan earmarks 5 percent of its grant funds for the ELSI program, which is expected to be approximately $5 million. This schedule reflects the five-year plan for the program: the goal is to be in a position by FY 1995 to complement the completion of the genetic reference map of the human genome with a slate of policy options addressing the highest priority challenges the uses of that map will pose.

Achieving this goal will require a combination of ethical, legal, and social scientific analysis, policy development, and public education and discussion. This need opens opportunities to develop new models for the forms of scholarship that currently go under the banners of bioethics, technology assessment, science studies, or health policy. For example, by combining scholarship and outreach, program grantees are developing models that better incorporate public voices into the inquiry. By collaborating across projects and institutions, they are improving their studies' interdisciplinary robustness as well. The challenge for all these investigators, like that of their colleagues on the scientific side of the Human Genome Project, is to coordinate their work toward its practical goal without sacrificing the breadth of perspective and intellectual creativity that their multiple lines of independent inquiry provide.

The NCHGR's program on ethical, legal, and social issues has attracted a great deal of public, scholarly, and congressional attention as an innovation in federal science funding. The program is certainly an experiment for NIH, the success of which greatly depends on the ability of projects like this volume to inspire informed public discussion and sound social policy.

This book, the individual contributions that its authors have made, and the agenda for research that their collective efforts have produced, constitute promising early results for this experiment. In the year since this volume's originating workshop, the NIH and DOE have launched dedicated research initiatives in each of the areas that the workshop participants identified as highest priority: the introduction of new genetic tests into medical practice, privacy protection for individual genetic information, genetic discrimination by employers and insurance companies, and the cultural impact of genetics on our understanding of ourselves. Even specific recommendations from individual authors, such as LeRoy Walters's call for a national advisory commission on genetic testing and screening (Chapter 16), have already found receptive congressional ears. We take these developments as preliminary evidence for the "ELSI hypothesis": that combining scientific research funding with adequate support for complementary research in the social sciences and humanities will help our social policies about science to evolve in a well informed and authoritative way. The high standards that this initial collection of agenda-setting essays provides for the inquiry and dialogue to follow endorses the merits of doing our science in the real world.

Contributors

George J. Annas, J.D., M.P.H.
Edward Utley Professor of Health
  Law
Director, Law, Medicine and Ethics
  Program
Boston University Schools of
  Medicine and Public Health
Boston, MA

Arthur L. Caplan, Ph.D.
Professor and Director
Bio-Medical Ethics Center
University of Minnesota
Minneapolis, Minnesota

C. Thomas Caskey, M.D.
Professor and Director
Institute of Molecular Genetics
Baylor College of Medicine
Houston, Texas

Rebecca S. Eisenberg, J.D.
Professor of Law
Michigan Law School
Ann Arbor, Michigan

Sherman Elias, M.D.
Director, Division of Reproductive
  Genetics
Professor of Obstetrics and
  Gynecology
College of Medicine
University of Tennessee, Memphis
Memphis, Tennessee

Patricia A. King, J.D.
Professor of Law
Georgetown University Law Center
Washington, D.C.

Ruth Macklin, Ph.D.
Professor of Bioethics
Department of Epidemiology and
  Social Medicine
Albert Einstein College of Medicine
Bronx, New York

Victor A. McKusick, M.D.
University Professor of Medical
  Genetics
Johns Hopkins University
Baltimore, Maryland

Thomas H. Murray, Ph.D.
Professor and Director
Center for Bioethics
Case Western Reserve University
   School of Medicine
Cleveland, Ohio

Robert N. Proctor, Ph.D.
Professor of History
Penn State
State College, Pennsylvania

John A. Robertson, J.D.
Professor of Law
University of Texas Law School
Austin, Texas

Evelyne Shuster, Ph.D.
Clinical Ethicist
Department of Veterans Affairs
   Medical Center
Philadelphia, Pennsylvania

Joe Leigh Simpson, M.D.
Faculty Professor and Chairman
Department of Obstetrics and
   Gynecology
College of Medicine
University of Tennessee, Memphis
Memphis, Tennessee

James Sorenson, Ph.D.
Professor
University of North Carolina School
   of Public Health
Chapel Hill, North Carolina

Judith P. Swazey, Ph.D.
President
Acadia Institute Bar Harbor, Maine

LeRoy Walters, Ph.D.
Director
Kennedy Center for Bioethics
Georgetown University
Washington, D.C.

Ray White, Ph.D.
Howard Hughes Medical Institute
University of Utah School of
   Medicine
Salt Lake City, Utah

Acronyms

ASHG	American Society of Human Genetics
CEPH	Centre d'Etude du Polymorphism Humain
DHHS	Department of Health and Human Services
DOE	Department of Energy
ELSI	Ethical, Legal, and Social Issues (Working Group and Program)
GDB	Genome Data Base (Johns Hopkins University)
HGM	Human Gene Mapping Workshop
HLA	Human leukocyte antigen
HUGO	Human Genome Organization
Kb	Kilobase
NAS	National Academy of Sciences
NCHGR	National Center for Human Genome Research
NIH	National Institutes of Health
NRC	National Research Council
NSF	National Science Foundation
OTA	Office of Technology Assessment (U.S. Congress)
PCR	Polymerase chain reaction
RFLP	Restriction fragment length polymorphism
SBIF	Small Business Innovation Research
STS	Sequence-tagged sites
VNTRs	Variable number tandem repeats
YAC	Yeast artificial chromosome

I

The Human Genome Project

1

The Major Social Policy Issues
Raised by the Human Genome Project

George J. Annas and Sherman Elias

In Edward Albee's 1962 play *Who's Afraid of Virginia Woolf?* George (a historian) describes the agenda of modern biology to alter chromosomes:

> the genetic makeup of a sperm cell changed, reordered . . . to order, actually . . .
> for hair and eye color, stature, potency . . . I imagine . . . hairiness, features,
> health . . . and *mind.* Most important . . . Mind. All imbalances will be cor-
> rected, sifted out . . . propensity for various diseases will be gone, longevity
> assured. We will have a race of men . . . test-tube-bred . . . incubator-born . . .
> superb and sublime.

George's view of the future was sinister and threatening in the early 1960s. But today, these same sentiments seem quaint and almost anti-intellectual.[1] Mapping and sequencing the estimated three billion base pairs of the human genome (our 50,000 to 100,000 genes) is in, raising serious questions about either the project itself or the ultimate application of the potential knowledge is out.

The *Wall Street Journal* summarized the case for the Human Genome Project in early 1989 when it editorialized, "The techniques of gene identification, separation and splicing now allow us to discover the basic causes of ailments and, thus, to progress toward cures and even precursory treatments that might ward off the onset of illness ranging from cancer to heart disease and AIDS." All that is lacking "is a blueprint—a map of the human genome." Noting that some members of the European Parliament had suggested that

3

ethical questions should be answered *"before it proceeds,"* the *Journal* opined, "This, of course, is a formula for making no progress at all." The editorial concluded, "The Human Genome [Project] . . . may well invite attack from those who are fearful of or hostile to the future. It should also attract the active support of those willing to defend the future."[2]

The National Institutes of Health (NIH) created a Center for Human Genome Research, which issued a 1989 request for funding proposals to study "the ethical, social and legal issues that may arise from the application of knowledge gained as a result of the Human Genome [Project]." The initial announcement made it clear that such projects are to be about the "immense potential benefit to humankind" of the project and to focus on "the best way to ensure that the information is used in the most beneficial and responsible manner." James Watson, until April 1992 the head of the Genome Center, has also been the Genome Project's leading proponent, having said, among other things, that the project provides "an extraordinary potential for human betterment. . . . We can have at our disposal the ultimate tool for understanding ourselves at the molecular level. . . . The time to act is now." And, "How can we not do it? We used to think our fate is in our stars. Now we know, in large measure, our fate is in our genes."[3]

These statements are, of course, hyperbole, and careful planning, although necessary, may prove insufficient to anticipate potential problems. As James Watson wrote in 1967, while reflecting on his own early career in science: "Science seldom proceeds in the straightforward logical manner. . . . its steps are often very human events in which personalities and cultural traditions play major roles."[4]

Predicting the Future

The Human Genome Project has frequently been compared to both the Manhattan Project and the Apollo Moon Program. These comparisons generally have to do with the magnitude of the project, and "big biology" is clearly happy to have its own megaproject of a size formerly restricted to physicists and engineers. As *Time* magazine put it, "Like atomic energy, genetic engineering is an irresistible force that will not be wished or legislated away."[5] But the sheer size of these two other projects obscures the more important lessons to be learned from them. The Manhattan Project is familiar, but it still teaches us volumes about the science and the unforeseen impact of technological "advance." In late 1945, Robert Oppenheimer testified about the role of science in the development of the atomic bomb before the U.S. Congress:

> When you come right down to it, the reason that we did this job is because it was an organic necessity. If you are a scientist, you cannot stop such a thing. If

you are a scientist, you believe that it is good to find out how the world works; that it is good to find what the realities are; that it is good to turn over to mankind at large the greatest possible power to control the world. . . .[6]

The striking thing in Oppenheimer's testimony is his emphasis on the notion that science is unstoppable with the simultaneous insistence that its goal is *control* over nature, irreconcilable concepts that seem equally at the heart of the Human Genome Project. Of course, with the atomic bomb, control became illusory. The bomb, which carries with it the promise of the total annihilation of mankind, has made the nation-state ultimately unstable and at the mercy of every other nation with the bomb. Necessity has forced all nuclear powers to move, however slowly, toward a transnational community.

The Apollo Project had its own problems. An engineering exercise, it was about neither the inevitability of scientific advance nor the control of nature. Instead, it was about military advantage and commercialism, disguised as politics and hyped as a peace mission. As Walter McDougall has persuasively documented, the plaque Astronaut Neil Armstrong left on the moon, which read, "We came in peace for all mankind," was ironic. In McDougall's words:

> The moon was not what space was all about. It was about science, sometimes spectacular science, but mostly about spy satellites, and comsats, and other orbital systems for military and commercial advantage. "Space for peace" could no more be engineered than social harmony, and the UN Outer Space Treaty . . . drew many nations into the hunt for advantage, not integration, through spaceflight.[7]

The *Wall Street Journal* seems more attuned to the commercial applications of gene mapping and sequencing than the NIH, although congressional support of the project is based primarily on the hope that the United States can maintain its lead in the biotechnology industries. Neither ethicists nor social planners played any real role in the Manhattan Project or the Apollo Project. They will play at least some minor role in the Genome Project, but how should that role be structured, and what should it be?

Genetic Precedent

In 1983, the President's Commission on Bioethics predicted that the cystic fibrosis gene was likely to be identified within the coming decade. Accordingly, the Commission recommended that we should begin to develop social and legal policies that would help determine who should be screened, what they should be told, how the information would be kept confidential, what the role of health insurers and employers should be, and what type of counseling and treatment should be available.[8] The gene was identified and screen-

ing tests made available in late 1989, but none of these issues had been seriously addressed in the intervening six years. Social policy on cystic fibrosis
screening (discussed in Chapter 11) is now being developed in a reactive
atmosphere that ensures that the wide-scale use of the test will be delayed, and
that potentially damaging mistakes are likely as some physicians begin routinely offering CF screening while NIH-sponsored pilot studies are under way.

The Social Policy Issues

Can we do better? Can social policy be developed either prior to or *simultaneously* with scientific research and technological advance? The U.S.
National Institutes of Health believes we can and have announced that as
much as 5 percent of their genome budget might be spent for study of the
"ethical, social and legal issues that may arise from the application of knowledge gained from the Human Genome Project." The Department of Energy
has concurred.

This approach is unique and could provide a model for other federally
financed research. It is crucial that all involved in it understand what is at
stake. Success in dealing with the ethical, legal, and social policy issues raised
by the Human Genome Project will require the active cooperation of the scientists engaged in the genome initiative. Many scientists may be skeptical that
this amount of funding should be devoted to this unprecedented effort. The
purpose of addressing these issues now is to help assure that society reaps a
maximum of benefit and a minimum of harm from the fruits of the genome
initiative.

There are many ways to summarize the ethical, legal, and social policy
issues raised by the Human Genome Project. NIH's National Center for
Human Genome Research initial request for proposals simply listed them in
a series of sentences, numbered one through nine, in no particular order. We
have found it more conceptually useful to roughly divide the issues into three
levels, which we designate individual/family, societal, and species. In describing the issues presented on each level, we will also identify those on which a
reasonable societal consensus has already been reached and those that remain
problematic or controversial. It should be noted at the outset that neither we
nor the contributors to this book have identified any uniquely novel social,
ethical, or legal issues raised by the Human Genome Project. On the other
hand, the medical and informational applications of the fruits of the Human
Genome Project mean that some of the issues previously dealt with in one or
a handful of genetic screening tests, for example, will have to be dealt with in
the future with hundreds or thousands more potential tests. Technically

speaking, this is simply a difference in degree, but if society focuses on genetics to the diminution of environmental factors, it could translate into a difference in kind, as some of the following examples illustrate.

Level 1: Individual/Family Issues

Level 1 covers issues involved in the direct use of genetic information that are no different in kind than those we have encountered using genetic information for such conditions as Huntington disease, sickle cell disease, and Tay-Sachs disease.[9] These issues include what information can be collected, how, by whom, on whose authority, for what purpose, how and to whom the information is disclosed, ownership of the technology, and the consequences for the individual and family. These issues are most directly applicable to the medical uses of genetic information, such as who should be screened, when to screen, what type of treatment and counseling should be available prior to screening, who pays, who has access to the information, and what information can and should be used outside the medical screening and treatment arena (such as insurance companies, employers, education, and the military).

As there have been large-scale genetic screening and counseling programs for such conditions as Tay-Sachs disease, sickle cell disese, phenylketonuria (PKU), and neural tube defects, it might be supposed that the major social policy issues raised by such screening have been solved. This supposition is incorrect. Each genetic disease has had special characteristics that have posed different problems. For example, some diseases occur most frequently in specific racial or ethnic groups, raising potential issues of discrimination and stigmatization. Other screening tests, such as those for neural tube defects, can be done only on pregnant women, and abortion is the only "treatment." Still others can be performed only on newborns, and screening for conditions such as PKU that require immediate treatment to prevent harm has been made mandatory by almost all states.

Two major ethical and legal issues are implicit in all genetic screening programs: autonomy and confidentiality. Autonomy requires that all screening programs be voluntary, and that consent to them is sought only after full information concerning the implications of a positive finding is both disclosed and understood. As the number of tests available proliferates, some form of "generic" genetic counseling prior to testing will have to be developed. Confidentiality requires that the findings not be disclosed to anyone else without the individual's consent. Although not a genetic disease, human immunodeficiency virus infection has provided an opportunity to see how widespread discrimination against individuals with a particular condition demands that testing be voluntary and that the results be kept confidential to protect the rights of individuals. Just how the Americans with Disabilities Act, which begins to take effect in 1992, will impact on genetic disabilities must also be analyzed and appropriate regulations or amendments adopted.

In the employment setting it has already been suggested that five principles should guide legislators, regulators, and professional groups in setting guidelines for medical screening: (1) medical inquiries of employees should be limited to job-related information; (2) only tests that are safe and of proven efficacy should be used; (3) applicants and employees should be informed of all medical tests in advance, given the results, and told when any employment decision will be based on test results; (4) intracompany and extracompany disclosure of medical records must be controlled and confidentiality assured; and (5) comprehensive, consistent, and predictable handicap discrimination legislation should be enacted.[10]

We suggest supplementing these worthy principles with three others directed at employers: (1) ethical issues involving screening should be fully explored *before* a screening program commences; (2) screening should only be done on an individual with the individual's informed consent; and (3) counseling should be available both before and after screening, and the resources for any reasonable intervention that can benefit the individuals screened should be in place and available to him or her before screening is offered. Legislation on both the federal and state levels will be required to protect confidentiality and minimize discrimination. In late 1990, Representative John Conyers (D., Mich.) introduced legislation to help protect the confidentiality of genetic information held by governmental agencies, including law enforcement agencies, and subcontractors of the federal government. This proposal is an essential first step, but only begins to provide the protection that ultimately will be needed.

As our current experience with cystic fibrosis screening illustrates, the ethical, social, and legal problems that will arise from the first level of concern are problems we have dealt with before in similar contexts and ones we are likely to treat similarly. Research at this level of concern is likely to prove most fruitful if it concentrates on analyzing the successes and failures of current and past genetic screening and counseling efforts (especially our experience with sickle cell disease, Tay-Sachs disease, neural tube defects, and Down syndrome). Special attention should also be devoted to those areas that traditionally have been ignored: public education and public financing of universal access—although neither of these social policy issues is unique to the Human Genome Project.

One problem that deserves more attention than it has received is the issue of being able to identify a disease without having any treatment for it long in advance of the disease's appearance. If Huntington disease is a reasonable model, great care needs to be taken in counseling individuals prior to offering presymptomatic testing of individuals. No one has yet seriously suggested population screening (or even family screening) for this condition, but there are likely to be calls for mandatory neonatal screening for conditions that could be aggravated by the environment (such as a propensity to develop skin cancer that could be avoided by avoiding exposure to the sun). This will not only raise issues of possible discrimination, stigmatization, and pathology, it will also raise issues of the need for parental consent for screening, and issues of child abuse or neglect for failure to comply with medical recommendations to prevent the disease from occurring. Standards for DNA "banks," including criminal banks for DNA identification, and access to the information about individuals they contain will also have to be developed and enforced. This is especially important because such banks will contain information that can only be decoded in the future, somewhat like a future diary, so that no one can know what information is stored. It is not too early to recommend the following minimum standards for the establishment of a DNA "gene bank":

1. No DNA bank should be created or begin to store samples until:
    a. public notice is issued, stating that the DNA bank is to be established, including the reason for the bank;
    b. a privacy impact statement is prepared and filed with a designated public agency that is also responsible for developing and enforcing privacy guidelines for the DNA bank; and
    c. the burden of proof should be on the DNA bank to establish that storage of DNA molecules is necessary to achieve an important medical or societal goal.
2. No collection of DNA samples destined for storage is permissible without prior written informed consent that:
    a. sets forth the purpose of the storage;
    b. sets forth all uses that will be permitted of the DNA sample;
    c. guarantees the individual continued access to the sample and all records about the sample, and the absolute right to order the identifiable sample destroyed at any time.
    d. guarantees the destruction of the sample (or its return to the individual) should the DNA bank significantly change its identity or cease operation.
3. DNA samples can only be used for the purposes for which they are collected. Specifically there may be:
    a. no waivers or "boilerplate" statements that permit other uses;
    b. no access to the DNA information by any third party without written notification to the individual;
    c. no access by third parties to any personally identifiable information;

    d. strict security measures, including criminal penalties for misuse or unauthorized use of DNA information.
4. Mechanisms must be developed to notify and counsel those whose DNA samples are in storage when new information that can have a significant health impact on the individual is obtainable from their stored DNA sample.

Every new *prenatal* test will, of course, raise the issue of whether the condition tested for is serious enough to justify terminating a pregnancy. For example, as the possibility of finding a treatment for cystic fibrosis grows, it is less and less justifiable to abort a fetus with this condition. Testing individuals for specific genes is a clear level 1 concern; genetic screening of groups or populations bridges levels 1 and 2 because screening is a public health endeavor that actively seeks out asymptomatic individuals, many of whom would not otherwise seek medical care or discover their condition.

We have historically viewed genetic testing and counseling as tools that permit individuals and families to use them as they see fit. This approach has followed the "medical model" of the beneficent doctor-patient relationship, which is a model of mutual consent in which decisions are made for the benefit of the patient consistent with the patient's values. This model has served us well, although it is coming under intense pressure as the costs of medical care continue to escalate and large numbers of citizens lack reasonable access to medical care. Also, as the U.S. Supreme Court continues to restrict and diminish our constitutional right of privacy, just what information the government can require from its citizens becomes more uncertain.[11]

Level 2: Societal Issues

Societal issues involved in the genome cluster in three areas: population-based screening, resource allocation and commercialism, and eugenics. Population-based screening can help eliminate a genetic condition or simply identify the incidence of a genetic condition in a population or the presence of a genetic condition in an applicant for a particular benefit (employment, insurance, immigration). Autonomy and confidentiality are the major legal issues involved, and population-based screening becomes socially problematic primarily when it is mandatory and the results are made known to others without consent. The other two areas involving societal issues are more uniquely societal.

The issue of resource allocation has at least three aspects. The first is the most obvious: What percentage of the nation's biology research budget should be devoted to the Human Genome Project? Answering this question requires us to consider how research priorities are set in science, and who should set them. Should Congress appropriate funds directly for the Human Genome Project (as it is currently doing), or should the program compete directly with other proposed research projects and be peer-reviewed?

The second aspect involves making the benefits of the Human Genome Project available to all those who want or need them. This raises at least two questions. The first is the issue of commercialism: Who "owns" and can patent the products that are produced by the Genome Project? Because much of this research is federally funded, should its benefits be in the public domain? Or should individual companies and scientists continue to be able to patent genetic sequences in order to encourage research? (See Chapter 14 for a full discussion of these issues.) The second question is: Should the genetic tests and their follow-up procedures be made part of a "minimum benefit package" under national health insurance (or some other scheme for universal access), or should they be available only to those people who can pay for them with private funds? Society must, of course, confront these questions with every new medical technology.

The third aspect of the resource allocation issue is probably the most intrinsically interesting. It involves determining the balance of resource priorities between the amount spent on identifying and treating genetic diseases and the amount spent directly on other conditions that cause disease, such as poverty, drug and alcohol addiction, lack of housing, poor education, and lack of access to decent medical care.

Could the fact that we are vigorously pursuing this project lead us to deemphasize environmental pollution, work site hazards, and other major social problems that cause disease, because we someday hope to find a "genetic fix" that would permit humans to "cope" with these unhealthy conditions? Given the fact that many of the homeless are mentally ill, with conditions that may be genetically determined and treatable, it has even been strangely and unpersuasively suggested that the fruits of the Human Genome Project may help solve society's homelessness and crime problem.[12]

The third level 2 issue, and the most important one, concerns eugenics. As discussed in detail in Chapter 4, this issue is perhaps the most difficult to address because of the highly emotional reaction many individuals have to even the mention of the Nazis' racist genocide, which was based on a eugenic program founded on a theory of racial hygiene.[13] Although repugnant, the Nazi experience demands careful study to determine what led to it, why scientists and physicians supported it and collaborated in developing its theory and in making possible its execution, and how it was implemented. In this regard, our own national experience with racism, sterilization, and immigration quotas will have to be reexamined. In so doing, we are likely to rediscover the powerful role of economics in driving our own views of evolution (in the form of social Darwinism) and who should propagate.

The U.S. Supreme Court, for example, wrote in 1927, with clear reference to World War I, that eugenics by involuntary sterilization of the mentally retarded was constitutionally acceptable based on utilitarianism:

> We have seen more than once that the public welfare may call upon the best citizens for their lives. It would be strange if it could not call upon those who already sap the strength of the State for these lesser sacrifices, often not felt to be such by those concerned, in order to prevent our being swamped with incompetence. It is better for all the world, if instead of waiting to execute degenerate offspring for crime, or to let them starve for their imbecility, society can prevent those who are manifestly unfit from continuing their kind.[14]

That may seem ancient history, but in 1988 the U.S. Congress's Office of Technology Assessment (OTA) in discussing the "Social and Ethical Considerations" raised by the Human Genome Project, developed a similar theme:

> Human mating that proceeds without the use of genetic data about the risks of transmitting diseases will produce greater mortality and medical costs than if carriers of potentially deleterious genes are alerted to their status and encouraged to mate with noncarriers or to reproductive strategies.[15]

The likely primary reproductive strategy, mentioned only in passing in the report, will not be sterilization, but will be genetic screening of human embryos—already technically feasible, but not nearly to the extent possible once the genome is understood. Such screening need not be required; people will want it—even insist on it as their right. As the OTA notes, "New technologies for identifying traits and altering genes make it *possible for eugenic goals to be achieved through technological as opposed to social control.*"[16]

The experiences of World War II were felt much more directly by Europeans than by Americans, so it should not be surprising that Europeans are much more vocal and articulate than Americans on this issue. At a Workshop on International Cooperation for the Human Genome Project held in Valencia in October 1988, it was suggested that the conferees agree on a moratorium on genetic manipulation of germline cells and a ban on gene transfer experiments in early embryos. The proposal, similar in spirit to that adopted at Asilomar in 1972, lost. This seems reasonable since such a moratorium is premature before the issues are publicly and vigorously debated. On the other hand, in October 1990 the new unified German government enacted an even wider ban on embryo research, embryo transfer, surrogate motherhood, and genetic experimentation, perhaps in acknowledgment of its past.

Much excellent work is underway on the history of medicine and its linkage to the eugenics movement. This scholarship should be made an integral part of any effort to understand the eugenics movement in the twentieth century. As discussed in Chapters 7 and 8, it will also be necessary to decide whether to use genetics to *improve* the species, and to articulate the philosophical and moral concerns that a change in the direction of genetics from prevention and treatment to enhancement and "improvement" would entail.

Level 3: Species Issues

Level 3 issues relate to the fact that powerful new technologies do not simply change what we can do; they as importantly change the way we think about ourselves. In this respect, maps may become particularly powerful thought transformers. Maps model reality and help us understand it. Copernicus and Vesalius published their great works in the same year, 1543. Copernicus's *On the Motions of Heavenly Bodies* made it clear that the earth rotated around the sun, not the other way around. The earth could no longer be seen as the "center" of the universe.

Vesalius's "maps" of human anatomy may have been even more important metaphors for us, for in dissecting the human body Vesalius insisted that human beings could nonetheless be understood only as whole beings, *human* beings rather than as parts that can be fitted together to manufacture life forms. As discussed in Chapter 6, this portrayal contrasts starkly with the bar graph illustrations used by contemporary geneticists in "mapping" the genome, which are devoid of human reference, as though they represented life without life. Does this reconceptualization of the human to a new "map" encourage us to travel into areas that could lead us to simultaneously misunderstand and demean what it is to be human?

What new human perspectives, or what new perspectives on humans, will a sequential map of the three billion base pairs of the human genome bring? The most obvious is that breaking "human beings" down into six billion "parts" is the ultimate in reductionism. The first concern is the consequence of viewing humans as an assemblage of molecules, arranged in a certain way. The almost inevitable tendency of such a view is expressed in *Brave New World.* People could view themselves and each other as products that can be "manufactured" and subject to quality control measures. People could be "made to measure," both literally and figuratively. If people are so seen, we might try not only to manipulate them as embryos and fetuses, but also see the resulting children as products themselves. This possibility raises the current stakes in the debates about frozen embryos and surrogate mothers to a new height—if children are seen as products, the purchase and sale of the resulting children themselves, not only embryos, may be seen as reasonable.

Second, to the extent that genes are inaccurately seen as more influential than environment, our actions may be viewed as "genetically determined," rather than resulting from free will. We have already witnessed an early example of this type of reasoning in the use of the "47,XYY defense." Individuals possessing the 47,XYY karyotype were thought to be more prone to commit crime. Accordingly, individuals accused of crime who also had an extra Y chromosome argued that they should not be held responsible for their actions because their genetic composition predisposed them to crime. This defense has been generally rejected, and in the few cases where it was accepted, the defendant was confined to a mental institution until "cured." Of course, since it is impossible to remove the extra Y chromosome from any cell, let alone every cell, in one's body, a cure is not possible.

In addition to potential use in the criminal law, perhaps in a misguided form of genetic screening followed by monitoring or "predelinquency detention," such genetic predispositions are likely to be used in education, job placement, and military assignments. For example, if intelligence in mathematics is found to be genetic, schools could use this information to track, grade, and promote the "genetically gifted" in math classes.

Finally, we know that most diseases and abnormalities are social constructs rather than facts of nature. Myopia, for example, is well accepted, but obesity is not. There is no "normal" or "standard" human genome that can be discovered. Nonetheless, we may invent one. If we do, what variations will society permit before an individual's genome is labeled "substandard" or "abnormal"? Moreover, what impact will such a construct of "genetic normalcy" have on society and on its members who are labeled "substandard"? For example, what variation in a fetus should prompt a couple to opt for abortion or a genetic counselor to suggest abortion? What variation should prompt a counselor to suggest sterilization? What interventions will society deem acceptable in an individual's life based on his or her genetic composition? Should health care insurance companies be able to disclaim financial responsibility for the medical needs of a child whose parents knew prior to conception or birth that the child would be born with a seriously "abnormal" genome? Should employers be able to screen out workers on the basis of their genomes?

These and many other similar issues based on screening for single-site genes exist today. But the magnitude of screening possibilities that may result from an analysis of the map of the human genome will raise these issues to new heights and will inevitably change the way we think about ourselves and about what it means to be human.

Developing Research Priorities

There is no magic blueprint for meaningful social policy input to scientific/ technological innovations. This makes the NIH/DOE effort even more important. If successful, it could help set a standard for future collaboration and public input into the social policy implications of research efforts before their results have been attained and society is forced to deal with their reality. Conversely, if unsuccessful, it may create the impression that such cooperative efforts are impossible and that science and society must remain distinct entities. To help prevent this latter possibility, we strongly recommend that the scientists involved in genome research take the time to explain the science and technology involved to philosophers, ethicists, lawyers, physicians, historians, and other social scientists who need to understand them because they will be suggesting or making social policy. It will also be necessary to keep an open mind to the possibility (we believe the reality) that careful ethical scholarship and public education and discussion *before* genetic tools reach the public will help maximize their benefit and minimize their harm, even though this discussion may delay actual development.

Since the Human Genome Project, at least as funded by the U.S. Congress, is an American undertaking, its characteristics are likely to be governed by American values and American culture. In this regard, for example, access to the products of the Human Genome Project will follow the patterns of access to other medical tests and treatments in the United States. This will mean that unless some form of universal health care access is established in our country in the future, the potential exists to exacerbate economic class differences (as discussed in Chapter 5) by making the products of the Human Genome Project available to those who can pay for them and denying them to the uninsured and the underinsured. Although the products of the Human Genome Project may lead us toward some form of national health insurance, focusing on this issue does not seem sufficiently unique to the Project to warrant much Genome Center–sponsored research.

Similar conclusions can be drawn with regard to patenting[17] (but see Chapter 14), commercialization, and population-based screening. Although this seems counterintuitive, the major focus of research efforts should probably be on level 1 and level 3 issues: level 1 because we have not yet determined how to make our commitments to autonomy and confidentiality a reality in today's world of private employers, private health insurance companies, and rampant discrimination based on all sorts of human differences; level 3 because these issues pose the greatest potential danger to our future. Special attention, we believe, must be paid to issues of reductionism and determinism, both of which could lead to an undermining of our concept of human dignity and a negation of our current views of human freedom.

In this regard, much more discussion, rather than simply research bans, should be focused on germline genetic therapy. To date, the arguments against it have been extremely weak and mainly religious-based, although the arguments for it (primarily efficiency) have been even weaker. Religious images, of course, have also been used to describe and sell the Human Genome Project itself, some calling it a search for the "Book of Man" and others the quest for the "Holy Grail" (See Chapter 3). Since manipulation of the human embryo may be the only way to actually influence a human's genetic composition or functioning, it is imperative that we vigorously and publicly confront and debate the pros and cons of taking this action before the technology arrives that make it possible.

It is still reasonable to hope that scientists and policy makers can enter the genetic age together, and with their eyes open to both the realistic benefits and the potential pitfalls of mapping our genes. The ethical, legal, and social policy issues facing the Human Genome Project are not equal. Priorities must be set and a research agenda worked out and agreed on, as a beginning step toward a responsible research program on the legal, ethical, and social issues raised by the Human Genome Project. It is the obligation of people who take legal and ethical issues seriously to ensure that the dangers, as well as the opportunities, are rigorously and publicly explored. We must get beyond Honey's response to George's musings on our genetic future in Albee's play: "How exciting!"

NOTES

1. Portions of this chapter are adapted from G. J. Annas, "Who's Afraid of the Human Genome?" *Hastings Center Report*, July 1989.
2. "Chromosome Cartography," *Wall Street Journal*, March 16, 1989, p. A16.
3. L. Jaroff, "The Gene Hunt," *Time*, March 20, 1989, pp. 62–7.
4. J. Watson, *The Double Helix*, Macmillan, New York, 1968, p. xi.
5. P. Elmer-Dewitt, "The Perils of Trading on Heredity," *Time*, March 20, 1989, p. 71.
6. R. Rhodes, *The Making of the Atomic Bomb*, Simon & Schuster, New York, 1986, p. 761.
7. W. McDougall, *The Heavens and the Earth*, Basic Books, New York, 1985, p. 413.

8. President's Commission, *Genetic Screening and Counseling,* Washington, D.C., 1983.

9. See generally S. Elias and G. J. Annas, *Reproductive Genetics and the Law,* Yearbook Medical Publishers, Chicago, 1987.

10. M. A. Rothstein, *Medical Screening and the Employee Health Cost Crisis,* Bureau of National Affairs, Washington, D.C., 1989.

11. N. A. Holtzman, *Proceed with Caution,* Johns Hopkins University Press, Baltimore, 1989.

12. D. E. Koshland, "Sequences and Consequences of the Human Genome," *Science* 246:189, 1989; and D. E. Koshland, "The Rational Approach to the Irrational," *Science* 250:189, 1990.

13. R. Proctor, *Racial Hygiene,* Harvard University Press, Cambridge, MA, 1987.

14. *Buck* v. *Bell,* 274 U.S. 200, 207 (1927).

15. U.S. Congress, Office of Technology Assessment, *Mapping Our Genes,* 1988, p. 84.

16. Ibid. (emphasis added).

17. The patenting issue may be more important, at least symbolically, than this volume treats it. It was, for example, the dispute over the wisdom of patenting cDNA fragments between NIH Director Bernadine Healy (she supports it) and James Watson (he calls it "sheer lunacy") that led to Watson's resignation as head of NIH's Center for Human Genome Research in April 1992. H. Stout, "Watson Resigns as Head of U.S. Gene-Mapping Project," *Wall Street Journal*, April 13, 1992, p. B9. We may also have undervalued somewhat the controversy over the use of "DNA fingerprinting" in the criminal courtroom, although this is ultimately a subset of the more generic and more central issue of the evidentiary standards courts should use in determining when a new scientific test should be permitted to be used by the jury to help them to decide the guilt of a defendant. See generally, National Research Council, *DNA Technology in Forensic Science,* National Academy Press, Washington, D.C., 1992.

2

The Human Genome Project: Plans, Status, and Applications in Biology and Medicine

Victor A. McKusick

The Human Genome Project aims to map and sequence completely the human genome. The sequence is the ultimate map. The National Research Council/National Academy of Sciences (NRC/NAS) committee, which concluded that the Human Genome Project was worthwhile and could be done in a reasonable time frame and budget, suggested "map first, sequence later."[1] No absolute stepwise dichotomy was in the mind of the committee, but this order was considered advisable because both the gene map and the "contig map" are necessary for most efficient sequencing, and the technology for sequencing requires improvement.

Given adequate funding (an estimated $200 million a year for the world-wide effort, in 1988 dollars), the Project should be completed by 2005. Spinoff in applications to medicine, in particular, will occur continuously during that period. The end product of the project is: a reference map (and sequence)—a source book for human biology and medicine for centuries to come. Two broad areas—variation and function—will occupy scientists for many years. It is urged by some that studies of variation in populations worldwide begin immediately because the disappearance of some populations through assimilation or attrition is erasing the history of evolution. Important as these studies are, it seems clear that the extent of variation is not such that confusion will arise in assembling the reference map. The journalist's question "Whose genome will be mapped and sequenced?" is irrelevant. Although the source of DNA used in sequencing should be recorded, the final map and sequence will be a composite of information from many individuals. The extent of variation of the human genome will need to be studied later. Variation and function are the essence of biology and genetics; it is ridiculous to expect the Human Genome Project to do it all and it would be imprudent to diffuse the effort. It will be important to "keep our eyes on the ball" if the main objective of a complete map and sequence is to be achieved in a timely manner and within the budget limits set.

Doing the entire job will represent economy of scale, and having the map/sequence information will facilitate greatly the elucidation of genetic disease. Cystic fibrosis is a case in point: identification of the gene was an expensive project and would undoubtedly have been much less expensive (how much less is also hard to say) if the complete sequence had been available.

Types of Maps

Gene mapping subsumes both genetic linkage maps, which are derived from meiotic recombination frequencies and measured in centimorgans (cMs), and physical maps, which are based on various experimental methods, several of which will be described later. The large-fragment clones that are produced by yeast artificial chromosome (YAC) cloning are permitting the creation of "contig maps" for each chromosome, that is, maps of overlapping, and therefore contiguous, DNA segments. Smaller segments, cosmid clones, have been used for the same purpose but are less efficient. These mapped segments can be used for mapping genes of unknown location by hybridization. They are also the raw material for nucleotide sequencing. The contig map is the penultimate physical map; the nucleotide sequence is the ultimate map.

The correlation of the genetic map (based on the location of cloned genes and anonymous DNA segments through family linkage studies) and the physical map (as represented by the contig map, for example) is likely to be aided by the use of sequence-tagged sites (STSs).[2] They may obviate, to a considerable extent, the need to store and distribute cloned DNA segments for study by investigators in many laboratories. If the STS identifier of a particular segment is known, the researcher can clone that segment and not require it as a clone from a repository. One will store information, not DNA, related to each part of the genome. STSs are a sort of Esperanto for scientists using diverse genetic and physical methods.

Finding all the genes could be helped by a concerted effort to create the cDNA map—the map of probes made by reverse transcription of messenger RNAs that have tissue and developmental stage specificity. The cDNA (or exon) map would provide candidate genes for diseases mapped by phenotype.

Since by definition the exons are "where the action is," sequencing could logically and efficiently start with them.

The Human Genome Project is getting underway against a background of human genetics research that has assembled at least some information on more than 5,000 of the 50,000 or more expressed genes in the human genome. The figure 5,000 comes from *Mendelian Inheritance in Man,*[3] a gene catalog that purports to carry one entry for each gene that has been characterized to any extent (Fig. 2–1). Furthermore, proposing to map all the genes, the project gets underway at a point where about 2,000 of the genes (4 percent or less) have been located to specific chromosomes and, in the case of most of them, to specific chromosomal regions (Fig. 2–2). It is useful to trace the history of this effort.

The History of Gene Mapping

The essence of Mendel's discovery in 1865 is that inheritance is particulate. The chromosomes were first described 12 years later by Walther Fleming, of

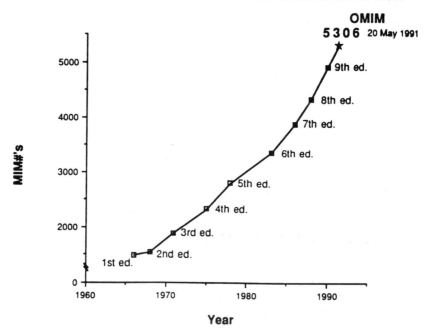

Figure 2–1. Total number of entries in successive editions of *Mendelian Inheritance in Man* and in the online version (OMIM).

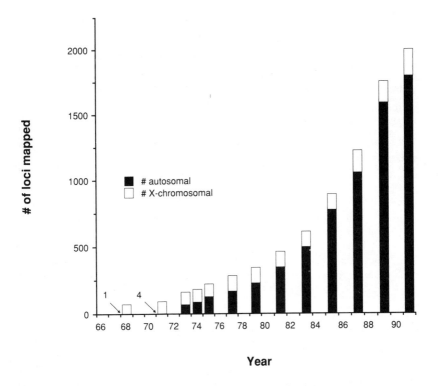

Figure 2–2. Growth of genes mapped to specific chromosomes.

Kiel. Meiosis, or the reduction division, as it was then called, was described in the 1880s. In part as the basis for rationalizing the reduction division of the chromosomes in gametogenesis, the notion was put forward that the chromosomes carry factors that determine development: the Roux-deVries-Weissmann hypothesis. Already before the rediscovery of Mendelism in 1900, E. B. Wilson was able to state in 1896:

> the chromosome is a congeries or colony of invisible self-propagating vital units . . . , each of which has the power of determining the development of a particular quality. Weismann conceives these units . . . [to be] associated in linear groups to form the . . . chromosomes.[4]

Thus, the chromosome theory of heredity and the concept of physical linkage of the genes preceded the rediscovery of Mendelism. The chromosome theory of Mendelism was advanced by Sutton and Boveri in 1902–1904.

The phenomenon of linkage was discovered in the first decade of this century in the domestic fowl by Bateson and Punnett, who introduced the terms "coupling" and "repulsion." The linear arrangement of Mendel's particulate

elements of heredity along the chromosome and the linkage method for estimating the intervals separating two such elements, by then called genes
(Johansson's term, about 1909), were conceived by Thomas Hunt Morgan
on the basis of studies in *Drosophila* beginning about 1911. Undergraduate
Alfred H. Sturtevant was important in developing the concept of linkage
mapping.

Also in 1911, E. B. Wilson, Morgan's colleague at Columbia, for the first
time assigned a specific gene to a specific chromosome in a mammal, the colorblindness gene to the human X chromosome:

> In the case of colorblindness, for example, all the facts seem to follow under this
> assumption [that the gene is on the X chromosome] if the male be digametic (as
> Guyer's observations show to be the case in man). For, in fertilization this char-
> acter will pass with the affected X chromosome from the male into the female,
> and from the female into half her offspring of both sexes. . . . Colorblindness,
> being a recessive character, should therefore appear in neither daughters nor
> granddaughters, but in half the grandsons, as seems actually to be the case.[5]

Friedrich Horner, a Zurich ophthalmologist, had described the typical pedigree pattern of color blindness in 1876.[6] As Horner pointed out, this pedigree
pattern was known also for hemophilia and later it was recognized for a few
other disorders, which, by the same reasoning as that applied by Wilson to
colorblindness, must also be coded by genes on the human X chromosome.

Because of the distinctive pedigree pattern, about thirty-six "sex-linked"
traits or disorders were described in the human before the first X-linked trait
was discovered in the mouse, about 1950. On the other hand, demonstration
of autosomal linkages proceeded much faster in the mouse because matings
such as the highly informative double-backcross mating could be done exper-
imentally in that species. The first linkage to be demonstrated in a mammal
was that between albinism and "pinkeye" in 1915 by the youthful J.B.S. Haldane and his sister Naomi, working with A. D. Sprunt. The authors gave the
following excuse for publishing a preliminary report: "Owing to the war it has
been necessary to publish prematurely, as unfortunately one of us (A.D.S.)
has already been killed in France."[7]

Although the interval separating the hemophilia and colorblindness loci on
the X chromosome was estimated by Haldane and his colleagues in two separate analyses in 1937 and 1947, no autosomal linkage was demonstrated, let
alone quantitated, until 1951—about the same time that the first X-linked
trait was demonstrated in the mouse. Jan Mohr, in his doctoral thesis in
Copenhagen, found linkage of Lutheran blood group with secretor factor. His
studies also suggested linkage of these two loci to that for myotonic dystrophy,

a finding that was subsequently confirmed. Indeed, the three loci were shown to be on chromosome 19, and it was mainly work in the department of Professor Jan Mohr, by then the long-time director of the Institute of Medical Genetics at Copenhagen, that provided the critical evidence of the chromosomal location of this linkage group.

In his studies in the early 1950s, Jan Mohr made use of the sib-pair method of Penrose, which is based on the principle that if two loci are linked, sibs will fail to show random association of traits determined by genes at those loci. The method of estimating the likelihood of linkage accounting for findings in particular pedigrees was developed by workers such as C.A.B. Smith and Newton Morton in the 1940s and 1950s. This work was the basis of the now familiar lod score—the logarithm of the odds of linkage as opposed to non-linkage.[8]

Between 1951 and 1968, the next watershed year, a number of further autosomal linkages were described, such as ABO vs nail-patella syndrome and Rh vs elliptocytosis-1. But for all these autosomal linkages, the precise chromosome carrying the genes was not known. In 1968, Roger Donahue, then a Ph.D. candidate in human genetics at Johns Hopkins University, demonstrated in his own family linkage of the Duffy blood group locus to chromosome 1 as distinctively marked in him and a number of his relatives by a so-called heteromorphism.[9] The relatively easy study of human chromosomes and the introduction of cytogenetics to the clinic had come in the previous ten years. In preparations made in 1968 in the just pre-banding era, Donahue's unusual chromosome had the appearance of an uncoiled area subjacent to the centromere. With the advent of banding methods about 1970, especially centromeric banding, it became clear that the anomaly represented an unusually long heterochromatic segment. Donahue had the wit to realize that this heteromorphism might be segregating as a dominant trait, since it was demonstrated by one of the two chromosomes No. 1, and could therefore be used as a linkage marker. Furthermore, he had the gumption to do a linkage study, which was not easy because his far-flung family made it difficult to collect blood samples and because of the labor involved in the testing of markers.

In the Donahue pedigree, the lod score for linkage of Duffy blood group to the long heterochromatic segment of chromosome 1 was far below the magic figure of 3.0 (1,000 to 1 odds on linkage) generally taken as evidence for linkage nowadays. However, the observation was quickly confirmed by others. Many studies of linkage of loci on chromosome 1, where now about 200 genes have been located (Table 2–1), indicate that the Duffy blood group is on the proximal short arm, i.e., on the opposite side of the centromere from the long heterochromatic segment with which it shows linkage. There may be sup-

Table 2–1. Number of Expressed
Genes Assigned to Each Chromosome
(Human Gene Mapping Workshop
10.5, Oxford, England, September 10,
1990)

Chromosome	Number of genes
1	194
2	110
3	68
4	74
5	72
6	104
7	107
8	53
9	59
10	58
11	132
12	100
13	24
14	60
15	51
16	68
17	107
18	23
19	91
20	34
21	36
22	58
X	187
Total	1868

pression of crossing-over in the pericentric region favoring demonstration of the linkage.

By 1968, when the first autosomal assignment was made, about sixty-eight genes had been assigned to the X chromosome on the basis of characteristic pedigree patterns. (The regional location was not known for any of the sixty-eight.) The growth of information on chromosomal assignments is diagrammed in Figure 2–1. As of mid-1991, about 2,000 expressed genes have been assigned to specific chromosomes and in most instances to specific bands of those chromosomes. This has been possible because of four commingling methodological streams: (1) family linkage study, (2) chromosome study, (3) somatic cell genetic study, and (4) molecular genetic study (Fig. 2–3).

Beginning about 1970, the pace of mapping accelerated, particularly through the use of somatic cell hybridization. The sorting out of the chromosomes in the subclones derived from rodent-human hybrid cells was com-

HUMAN GENE MAPPING
FOUR COMMINGLING METHODOLOGIC STREAMS

	Family Studies	Chromosome Studies	Somatic Cell Studies	Molecular Studies
FLS	Linkage (F) Linkage disequilibrium (LD) Ovarian tumor (OT)	Linkage with hetero- morphism or rearrangement (Fc) Deletion mapping (D) (qualitative)	Homology of synteny (H)	Linkage with RFLPs (Fd)
CH		Dosage effect (D) (quantitative) Exclusion mapping (EM) one form Chromosome aberration e.g. deletion (Ch) Virus-induced changes (V)	Assignment by SCH (S) Regional mapping by SCH (S) Chromosome-mediated gene transfer, CMGT (C)	*In situ* hybridization (A) Molecular analysis of flow sorted chromo- somes (REb)
S			SCH synteny test (S) Radiation induced gene segregation (R) Microcell-mediated gene transfer, MCGT (M)	DNA or RNA hybrid- ization in solution (HS) Southern analysis (REa) DNA-mediated gene transfer, DMGT (DM)
M				Restriction enzyme fine mapping (RE) DNA sequencing (NA) AA sequencing (Lepore approach) (AAS)

Figure 2–3. Four commingling methodological streams in human gene mapping.

parable to the assortment of human chromosomes in meiosis. Somatic cell hybridization was thus what Pontecorvo referred to as a "parasexual method" and Haldane called a "substitute for sex." The development of chromosome banding, also about 1970, provided an important, indeed essential, method-ological element by permitting the unique identification of each human chro-mosome and the differentiation of human chromosomes from rodent chro-mosomes in the hybrid cells.

About 1980, a further acceleration in gene mapping occurred with the introduction of molecular genetic techniques. Recombinant DNA technol-ogy provided probes that could be used in three ways: first, in connection with DNAs from somatic hybrid cells (obviating the necessity to have gene expres-sion in the cultured cells, since one could directly "go for the gene"); second, for in situ hybridization to chromosome spreads; and third, as DNA markers—for example, restriction fragment length polymorphisms (RFLPs) and variable number tandem repeats (VNTRs)—in family linkage studies.[10] Chromosomes sorted by means of the fluorescence-activated cell sorter could also be probed with the new armamentarium.*

The Status of the Human Gene Map

At the Human Gene Mapping Workshop in Oxford early in September 1990, the available information on the human gene map was collated, with results that are tabulated in Table 2–1 and Figure 2–4. Almost 1,900 genes had been mapped to specific chromosomal locations. In addition, over 4,500 DNA segments, so-called anonymous DNA segments because their function, if any, is not known (indeed most are known not to be expressed), had been mapped to specific chromosomal sites. About half these segments had been shown to be polymorphic, and many of them were sufficiently variable to make them useful as linkage markers.

The human is estimated to have about 50,000 genes, almost certainly not more than 100,000. Based on indirect arguments, estimates of this order were arrived at in the past; in more recent times, studies of the density of genes on the chromosomes and information on the range of sizes of genes support these estimates.

Many more genes have been assigned to some chromosomes than to others, if one examines either the absolute numbers (Fig. 2–4A) or the relative numbers, derived by dividing the absolute number by the length of the chromosome expressed as a percentage of the length of the haploid set of autosomes (Fig. 2–4B). The preponderance of loci on the X chromosome derives, of course, from the relative ease of assignment by pedigree pattern; this factor was aided by the availability of a selection system, based on the X-linked hypoxanthine phosphoribosyl transferase (HPRT) locus, that would be used in mapping by somatic cell hybridization. The large number on chromosome 17 is related in large part to the availability of the selection system based on the thymidine kinase locus on that chromosome. Chromosomes 11 and 16 are rather well mapped, perhaps particularly because of the location there of the beta-globin and alpha-globin loci, respectively. Chromosome 16 also had

---

*Chromosome sorting by this physical method serves the same role as meiosis in vivo and somatic cell hybrid clones in vitro. In situ hybridization and somatic cell hybridization are methods of physical mapping; family linkage studies give information on the genetic map, but when the markers used have been mapped to specific chromosomal sites, physical location can be inferred.

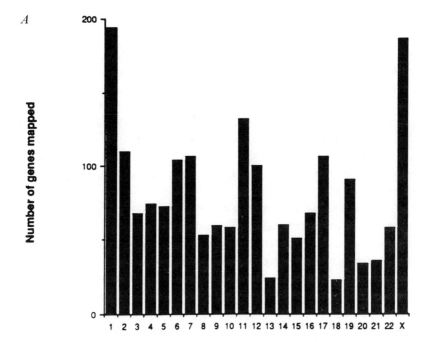

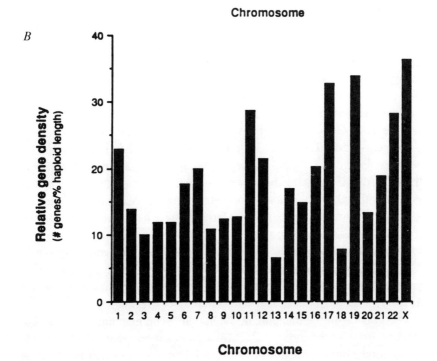

Figure 2–4. (A) Number of genes mapped to each chromosome. (B) Relative gene density (number of genes mapped to the chromosome/percent haploid length represented by that chromosome).

early on a good linkage marker in the form of the haptoglobin locus. Chromosome 6 had the advantage of HLA, a superb highly polymorphic marker system for genetic linkage. Chromosome 7 came in for intensive study when it was discovered that the cystic fibrosis locus maps to 7q. Chromosome 21 has had attention for the obvious reason of interest in Down syndrome. The strong performance of chromosome 22 may be related to the fact that the lambda light-chain genes are there, and the BCR/ABL fusion gene that results from the 9;22 translocation of chronic myeloid leukemia has stimulated much study.

The reason for the large number of genes assigned to chromosome 19 (the highest relative number second only to the X chromosome) may be merely that it got started early with the secretor/Lutheran/myotonic dystrophy linkage group, and the relatively large number on chromosome 1 may have the same explanation. But is it possible that some chromosomes are genically less densely populated than others? Chromosomes 13 and 18 stand out for a low number of mapped genes. It is noteworthy that except for the smallest autosome, No. 21, only No. 13 and No. 18 show trisomy that is compatible with live birth. This fact suggests also that there are relatively few genes, or at least few genes of critical importance, on these chromosomes.

That the relatively high density of genes on chromosome 19 ("the Java of the autosomes") and some others and the low density on chromosomes 13 and 18 is bona fide is supported by the studies of adenine-thymine (AT) or guanine-cytosine (GC) content of the chromosomes. Langlois and colleagues did dual-beam chromosome sorting after staining of the chromosomes with two dyes: chromomycin A3, a GC-specific dye, and Hoechst 33258, an AT-specific dye.[11] GC-rich parts of the genome (Giemsa-light bands) are known to be gene-rich; AT-rich portions (Giemsa-dark bands) are gene-poor. As shown in Figure 2–5, the chromosomes "off the diagonal" in the direction of the AT ordinate—Y (already known to have few genes), 13, and 18—by this evidence are relatively sparsely populated. Contrariwise, chromosomes 19, 17, 22, and some others, but particularly 19, have a relatively high concentration of genes.

Within chromosomes also, genes are, it seems, not uniformly distributed. More are found in Giemsa-negative bands than in Giemsa-positive bands, particularly in subtelomeric portions of the chromosomes. Examples include 1p15.5, 21q22.3, Xp22, and especially Xq28 (see Fig. 2–6).

In parallel with the mapping in the human, gene mapping has been proceeding apace in the mouse, which is the most useful mammalian genetic model for the human because of the extensive amount of genetic information already available and the relative ease of experimental manipulation. At the same time, June 1976, mapping in both the mouse and the human achieved the point that at least one gene had been assigned to every chromosome.

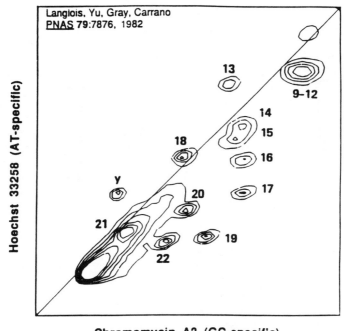

Langlois, Yu, Gray, Carrano
PNAS 79:7876, 1982

Hoechst 33258 (AT-specific)

13

9–12

14

15

18

16

y

20

17

21

19

22

Chromomycin A3 (GC-specific)

Figure 2–5. Distribution of chromosomes from dual-beam sorting using the two dyes indicated. (From Langlois, Yu, Gray, Carrano, *Proceedings of the National Academy of Sciences USA* 79:7876, 1982.)

Although in the mouse, as mentioned earlier, linkage mapping proceeded much faster until the surrogate methods such as somatic cell hybridization were developed, mapping of the human has outpaced mapping of the mouse in recent times.

Caenorhabditis elegans, a nematode, is also a useful model.[12] As in the human, there is extensive anatomical and physiological information, and, relatively speaking, the genetic and developmental information is even more extensive than in the human. The developmental lineage of each of the 959 cells of the organism is known; all the synapses of the nervous system have been described; more than 1,000 genes have been mapped to one or another of the four chromosome pairs. Contig maps are approaching completion. All that remains is nucleotide sequencing.

Methods of Gene Mapping, Old and New

As indicated in Table 2–2, somatic cell hybridization in all its variations has been by far the most productive method of human gene mapping. The

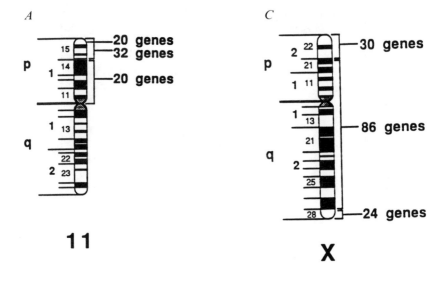

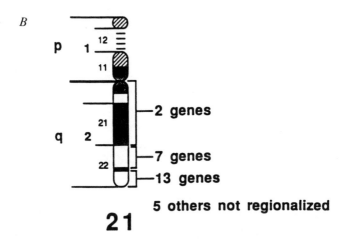

Figure 2–6. Examples of inhomogeneous distribution of mapped genes. (A) Chromosome 11, showing the highest density of genes in the terminal bands 11p15.5. (B) Chromosome 21, showing the highest density of genes in band 21q22.3. (C) Chromosome X, showing the highest density of genes in the terminal Giemsa-light bands of both arms.

Table 2-2. Number of Autosomal Loci Mapped by the Several Methods—May 15, 1991

Method	Number of loci mapped
Somatic cell hybridization	1215
In situ hybridization	744
Family linkage study	492
Restriction enzyme fine analysis	212
Dosage effect	162
Chromosome aberrations	124
Homology of synteny	121
Radiation-induced gene segregation	18
Others	146
Total	3234

method second in position is in situ hybridization, which attained this place by the fall of 1987. Its rapid ascendancy is in some ways surprising because the method was made to work reliably for single-copy genes only as recently as 1981. Presently, whenever a gene is cloned, one has not adequately characterized it until one has mapped it, first by determining its chromosomal location by hybridizing a gene probe to a panel of DNAs from somatic cell hybrids, and second by corroborating that assignment and regionalizing it by in situ hybridization. If a newly cloned gene is found to have an associated RFLP, one has a third method for mapping the gene: linkage against DNA markers that have previously been mapped in the CEPH families. The Centre d'Études du Polymorphisme Humain in Paris, developed by Nobel Laureate Jean Dausset, has a collection of cell lines from more than forty three-generation families in which all four grandparents and eight or more grandchildren, as well as their parents, have been available for sample collection. The DNAs from the families have been subjected to linkage studies to create a reference map for each chromosome. Nonisotopic labeling for in situ hybridization[13] has the advantage that one does not need to wait for development of the autoradiographs or have the nuisance and expense involved in the handling of radioisotopes. In addition, one can expect to study the relative position of two or more genes by using fluorescent markers of different color.

There are, however, many genes, including those of clinically important disorders such as cystic fibrosis, Huntington disease, and polycystic kidney disease, that could not be mapped by first cloning the gene because the nature of the mutant gene product was not known. In these cases, mapping of the phenotype by family linkage studies has been necessary. The availability of DNA markers such as RFLPs, distributed more or less uniformly over the genome at intervals of 5 to 10 cM to constitute a linkage reference map, means that the chances are good that linkage can be demonstrated between a given rare dominant (or even recessive) and one of the markers. The informativity of a particular RFLP depends on its degree of heterozygosity. On the suggestion of David Botstein and colleagues,[14] the informativity is measured by the polymorphism information content (PIC) value. PIC is the sum of the proportions of all parental matings that have at least one heterozygous parent. The disorders listed in Table 2–3 were all mapped by genetic linkage to DNA markers. The total in Table 2–2 adds up to much more than the total number of mapped autosomal loci because many of the genes have been mapped by more than one method.

Application of the polymerase chain reaction (PCR) to single sperm, by Arnheim and his colleagues,[15] permits the direct ascertainment of linkage and estimation of recombination frequencies. It is similar to determining directly the genetic composition of the "stick diagrams" used in textbooks to explain linkage and recombination and to present the data derived from family studies. The method has the potential advantage that one can study large numbers of meioses from a single doubly heterozygous male. Ordinarily, in human families, when one must depend on analysis of the phenotype of his offspring for identification of recombination, one has an opportunity to study no more than eight or ten or, in truly exceptional circumstances, sixteen or eighteen meioses from a single male. Because of the large number of meioses that can be studied by the Arnheim method, information on close linkage can be determined. Furthermore, the possibility of differences in the frequency of recombination in the same DNA segment in different males, as well as the effect of age in the individual male, can be studied.

Linkage disequilibrium can be used as a clue to close linkage. In a sense, homozygosity mapping of recessive genes is based on this principle.[16] In populations such as the Amish or the Finns where founder effect makes it likely that all cases of a given, ordinarily rare recessive are descendants from a common ancestor, one expects that genes closely situated in the disease-producing allele will stand a good chance of being transmitted to affected descendants and not to the unaffected. The closer the genes, the greater the chance that both the marker and the disease allele will remain on the same chromosome in a majority of affected persons. This is the argument underlying the use of haplotypes to identify the presence of thalassemia or other disease gene.* In the Amish, because of the defined genealogies, DNA panels that reflect an identifiable number and location of meioses might be used in the same way the reference families in CEPH are used.[17]

---

*Haplotype refers to the combination of specific alleles at several closely linked loci, for example, the haplotypes at the major histocompatibility complex for HLA types, such as A3, B27, C2, DR2. The approach was adapted to tracing beta-thalassemias and sickle cell anemia in the early 1980s and more recently has been used in the study of inborn errors of metabolism such as phenylketonuria and cystic fibrosis.

Table 2–3. Mapped Mendelian Disorders for Which the Biochemical Basis Was Not Previously Known (Mapping as a First Step Toward Basic Understanding)

Charcot-Marie-Tooth disease	1q,17,Xq13
Usher syndrome	1q
van der Woude lip-pit syndrome	1q
Aniridia	2p, 11p
Waardenburg syndrome	2q
von Hippel-Lindau syndrome	3p
Huntington disease	4p
Facioscapulohumeral muscular dystrophy	4q
Spinal muscular atrophy, several types	5q
Adenomatous polyposis of colon	5q
Hemochromatosis	6p
Juvenile myoclonic epilepsy	6p
Spinocerebellar ataxia (1 form)	6p
Craniosynostosis	7p
Greig craniopolysyndactyly syndrome	7p
Cystic fibrosis	7q
Langer-Giedion syndrome	8q
Friedreich ataxia	9q
Torsion dystonia	9q
Tuberous sclerosis	9q, 11q
Nail-patella syndrome	9q
Multiple endocrine neoplasia, type II	10q
Long QT syndrome	11p
Wilms tumor-1	11p
Multiple endocrine neoplasia, type I	11q
Ataxia-telangiectasia	11q
Retinoblastoma	13q
Wilson disease	13q
Marfan syndrome	15q
Polycystic kidney disease	16p
Batten disease	16p
Cataract, Marner type	16q
Neurofibromatosis	17q
Myotonic dystrophy	19q
Malignant hyperthermia	19q
Alzheimer disease (1 form)	21q
Amyotrophic lateral sclerosis (1 form)	21q
Acoustic neuroma, bilateral	22q
Duchenne muscular dystrophy	Xp
Hypophosphatemia, X-linked	Xp
Retinitis pigmentosa (2 forms)	Xp
Wiskott-Aldrich syndrome	Xp
Alport syndrome	Xq
Cleft palate, X-linked	Xq

In the 1970s Goss and Harris developed a method for determining the interval between genes on the basis of the chance that they would be separated when a standard dose of radiation was administered to the cell.[18] The radiated cell was "rescued" by hybridization with a rodent cell. This method of radiation-induced gene segregation, or radiation mapping, has been developed further by David Cox and others.[19] The product might be called the "zap map," a term Cox dislikes.

In recent times we have learned how to use enzymes that cut the DNA only rarely, at intervals of a few hundred kilobases or more. In general, the larger the number of nucleotides represented by the recognition site for the restriction enzyme, the rarer is the cut site and the larger the fragments produced. In addition, the methods for separating large fragments, of which the Schwartz and Charles Cantor method of pulsed-field gel electrophoresis (PFGE) was a pioneer, provide the material for hybridization of gene probes. Given that gene A has been assigned to a particular location in the genome by some other method, if gene B hybridizes to the same large fragment, one has determined where gene B is located also, as well as placed an upper limit on the interval separating genes A and B. Restriction mapping using frequently cutting endonucleases can define the relationship of A and B in more detail. YAC cloning also provides large DNA fragments to which cloned genes can be hybridized for mapping purposes.

The Sociology of Human Gene Mapping

Information on the human gene map has been collated in a series of human gene mapping (HGM) workshops.* Beginning with the Oxford workshop, the

*The first of these was convened at Yale by Frank Ruddle in 1973, with the financial support of the March of Dimes, represented by Daniel Bergsma, Vice President for Professional Education. Subsequent workshops were held in The Netherlands (D. Bootsma, organizer), Baltimore (V. A. McKusick, organizer), Oslo (K. Berg, organizer), Los Angeles (R. Sparkes, organizer), Helsinki (A. de la Chapelle, organizer), Paris (J. Frezal, organizer), and again New Haven (F. Ruddle and K. Kidd, co-organizers). The workshop in early September 1990 was held in Oxford, England, under the direction of Sir Walter Bodmer and Ian Craig; HGM11 took place in London in August 1991, under the direction of Sir Walter Bodmer and Ellen Solomon. For these HGM workshops a committee with cochairs assumes responsibility for collating the information on the gene map of each chromosome. A standing committee of the HGM workshops, cochaired by Phyllis J. McAlpine and Thomas B. Shows, renders opinions on matters of nomenclature, particularly choice of gene symbols, to assure consistency. Committees at the workshops are responsible for other overall considerations, namely, a committee on clinical disorders that have been mapped and a committee on neoplasia that reports on specific chromosomal changes associated with specific neoplasms, as well as the mapping of oncogenes and genes at cloned breakpoints in chromosome rearrangements in neoplasia and, recently, loss of heterozygosity and coamplification of genes in neoplasms as indicators of map location.

data on the human gene map are entered directly into a database, which will be continuously updated. This database, maintained in Baltimore at Johns Hopkins University, is called GDB (for genome database). Its development was funded by the Howard Hughes Medical Institute; since September 1, 1991, it has been supported jointly by the genome programs of the NIH and DOE, with supplemental support from several foreign sources. The chairs of each of the committees will have responsibility for ongoing editing, which they can perform from remote sites. The database will, furthermore, be generally accessible by the scientific community worldwide. Its development is under the direction of Peter Pearson, formerly chairman of the Department of Human Genetics of Leiden University and a leading cytogeneticist and gene mapper.

Fully integrated with GDB and distributed worldwide along with it (and indeed a main reason for the development of GDB in Baltimore) is OMIM, the online version of *Mendelian Inheritance in Man,* the encyclopedic catalog of genes that I have maintained on computer since late 1963 and have published as a computer-based book since 1966.[20] My colleagues and I have had experience with providing the scientific and medical community access to the online version for more than three years. Although the print version has appeared at two-year intervals, in recent times the rapid evolution of the field gives the online version the extra value of timeliness. Another advantage of the online version is the ability to search the database electronically. The size of the database is reflected by the number of entries, more than 5,000 (Fig. 2–1), each representing a single gene locus, as well as by the counts of individual authors (more than 57,000) and literature citations (more than 37,000). The database is clearly too large to be handled efficiently by print indices.*

The Human Genome Organization has designated GDB as the official repository for data generated by the Human Genome Project, short of bulk DNA sequence data. HUGO is the coordinating organization for the worldwide effort[21]; the Human Gene Mapping Workshops have become a component of HUGO.

It is anticipated that the worldwide effort to map and sequence the entire human genome will be organized on a chromosome-by-chromosome basis; that consortia of laboratories and investigators in many countries will be

---

*Within the United States, OMIM and, more recently, GDB have been easily accessible through Telenet and Internet; access and even transcription charges have been waived through the generosity of the Howard Hughes Medical Institute. To facilitate distribution elsewhere in the world, arrangements have been made for distribution nodes in the United Kingdom, Germany, Sweden, and Japan. For establishing access, contact GDB/OMIM User Support, Welch Medical Library, Johns Hopkins University, 1830 E. Monument St., Baltimore, MD 29205, USA. Tel.: (301) 955-7058; FAX: (301) 955-0054.

involved for each chromosome; and that these investigators will have frequent meetings for exchange of information—these meetings will largely replace the annual or biennial, increasingly large HGM workshop meetings. The chromosome-specific consortia will be administered by HUGO, will ensure that information gets into GDB in a timely and accurate manner, and, in the long run, will see that the job gets done.

Figure 2-1 indicates the growth in number of entries in MIM over the 25 years of its existence. Only one entry is created for each gene locus. All the allelic mutations at a given locus are listed under a single entry. Thus, the total gives some indication of the proportion of all genes about which we have information. That proportion is now about 10 percent. Until about 1970, almost the only method for identifying genes was the occurrence of Mendelian variation—hence the title *Mendelian Inheritance in Man.* Since about the 1970s, somatic cell hybridization provided a substitute for Mendelism. Many genes, such as that for thymidine kinase, were delineated through assignment to specific chromosomal locations by study of interspecific (for example, mouse-human) somatic cell hybrids. Since 1980 and the advent of molecular genetics, many genes have been cloned, mapped, and sequenced without any Mendelian variation in phenotype having been associated with them. All such genes, delineated and mapped by non-Mendelian, parasexual methods, have deservedly been incorporated as entries in MIM. These have contributed greatly to the recently accelerated pace of accessions.

The Role of Gene Mapping in Human Biology and Medicine

As Sir Walter Bodmer pointed out, what seemed like a rather recondite activity when the human gene mapping workshops were initiated in 1973 has achieved a central role in both human biology and scientific medicine. Charles Scriver of Montreal suggested that gene mapping is providing a neo-Vesalian basis for medicine. Beginning in the 1950s, the availability of relatively easy methods for microscopic study of the human chromosomes gave the clinical geneticist "his organ" and abetted the development of human genetics as a clinical specialty. Continually improving methods of chromosome study, particularly the methods for mapping genes on chromosomes, have given all of medicine a new paradigm. Specialists in all medical areas approach the study of their most puzzling diseases by first mapping the genes responsible for them. Thus, just as Vesalius's anatomical text of 1543 formed the basis for the physiology of William Harvey (1628) and the morbid anatomy of Morgagni (1761), gene mapping is having a widely pervasive influence on medicine.

In examining the significance of mapping information in clinical medicine, it may be useful to substitute the anatomical metaphor ("the anatomy of the human genome") for the cartographic metaphor ("the human gene map"). The anatomical metaphor prompts one to think in terms of the morbid anatomy, the comparative anatomy and evolution, the functional anatomy, the developmental anatomy, and the applied anatomy of the human genome.

The approximately 2,000 expressed genes that have been mapped to specific chromosomes, and in most instances to specific chromosome regions or bands (Table 2–1), code for blood groups, enzymes, hormones, clotting factors, growth factors, receptors (for example, for hormones and growth factors), cytokines, oncogenes, structural proteins (for example, collagens and elastin), and so forth. They also code for a large number of disease genes for which the biochemical basis is not known or was not known before the mapping (Table 2–3).

In all, the chromosomal location of more than 500 genetic disorders ("the morbid anatomy of the human genome") has been determined. Said differently, and perhaps more accurately, at about 500 of the 2,000 mapped loci, at least one disease-producing mutation has been identified. In the case of some disorders, the mapping has been done by locating the "wild-type" gene, such as that for phenylalanine hydroxylase, which is deficient in PKU and maps to 12q. In other disorders, the Mendelizing clinical phenotype has been mapped by family linkage studies using anchor markers such as RFLPs; examples are Huntington disease, polycystic kidney disease, and Marfan syndrome. In yet other disorders, mapping has been done by both approaches; for example, elliptocytosis, type I, was mapped to the distal region of the short arm of chromosome 1 by linkage to Rh and other markers situated there and also by mapping of the gene for protein 4.1 (which is mutant in that disorder) to the same region.

To date, the *applied anatomy* of the human genome has related particularly to those disorders for which the basic biochemical defect was not yet known (Fig. 2–7). Because of this "ignorance," it was impossible to devise a fully specific diagnostic test or to design rational therapy. Once the chromosomal location of a disease-producing gene is known, together with its proximity to other genes and especially DNA markers, one can make a diagnosis (prenatal, presymptomatic, and carrier) by the linkage principle. Furthermore, one can expect to determine the fundamental nature of the genetic lesion by "walking" or "jumping" in on the gene, a process often labeled, with questionable appropriateness, reverse genetics. Then, knowing the nature of

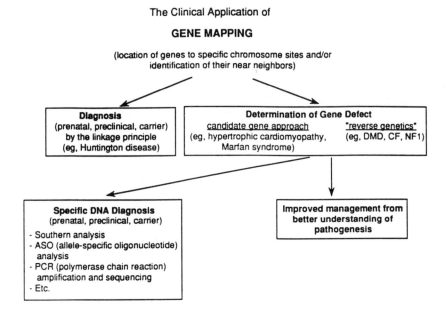

Figure 2–7. The clinical application of gene mapping.

the wild-type gene, one can work out the pathogenetic steps that connect gene to phene, mutation to clinical disorder. Secondary prevention and therapy through modification of those steps can be developed. In many instances gene therapy will probably find gene mapping information useful background.*

In connection with the *morbid anatomy* of the human genome, mapping, perhaps more than any other single factor, has been responsible for the establishment of the chromosome basis of cancer: by the demonstration of specific, microscopically evident chromosomal aberrations in association with specific neoplasia, by the mapping of oncogenes and antioncogenes (recessive tumor suppressor genes), and by the correlation of the two sets of observations. Somatic cell genetic disease, as the basis not only of all neoplasia but also of some congenital malformations and probably of autoimmune dis-

---

*Walter Gilbert,[22] as well as others, pointed out that going from the phenotype to the gene is not different from the process in classical genetics, for example in *Drosophila,* and therefore should not be described as "reverse." He suggested that true reverse genetics will be the increasing practice of the future when segments of the human genome that have been sequenced are studied for their functional significance. The process of going from the gene to the phenotype will be true "reverse genetics." As a substitute for "reverse genetics," in designating the process of cloning genes from their chromosomal location, Francis Collins uses the term "positional cloning," as opposed to "functional cloning," the process that involved starting with the protein gene product or mRNA.

eases, joins the three cardinal categories of disease as to genetic factors: single-gene disorders, multifactorial disorders, and chromosomal disorders.

Mitochondrial genetic disease is a fifth category. The mitochondrial chromosome, the twenty-fifth chromosome, was completely sequenced by 1981 in the laboratory of Fred Sanger at Cambridge, and its genes were mapped in the six years that followed. Deletions and point mutations in the mitochondrial chromosome were then identified as the basis of Leber optic atrophy, myotonic epilepsy/ragged red fiber disease, Pearson pancreas–bone marrow syndrome, oncocytomas, and, in cultured cells, chloramphenicol resistance. This sequence of discovery is the opposite, for the most part, of that pursued to date in the delineation of the nuclear genome, where the progression has been from study of diseases, then to their mapping, and finally to the sequencing of the genes. The mitochondrial chromosome, with its mere 16,569 base pairs, can be considered a paradigm for what the Human Genome Project hopes to achieve for the nuclear genome (which is approximately 200,000 times larger in terms of nucleotides). Indeed, complete sequencing may help greatly in identification of all the nuclear genes, just as it did in the case of the mitochondrial genes. At present, we have information, as cataloged in *Mendelian Inheritance in Man,* on more than 10 percent of the genes, or gene loci (Fig. 2-1). Walter Gilbert suggested that complete sequencing may be the best way to find the rest. Thereafter, one can determine the function of the genes and the disorders caused by mutations therein, as has been done for the mitochondrial chromosome. The cDNA map may be especially useful in "finding all the genes."

Counter-arguments and Counter-counter-arguments concerning the Human Genome Project

Among scientists, the main arguments against the Human Genome Project seem to be four: that it is bad science; that it is big science (with the implication that it is not the way science is most effectively done); that it creates an improper milieu for doctoral training in science; and that it is taking money away from other worthy (in the opinion of some, more worthy) projects.*

The purpose of the Human Genome Project is to create a tool for science—the source book referred to earlier. A good deal of applied science and engineering will go into the Genome Project, but it may not be appropriate to criticize the Project through a comparison with biological science as it has been pursued traditionally.

*I am indebted to Leroy E. Hood for this listing of counter-arguments and, in part, for the formulation of counter-counter-arguments.

The Human Genome Project is not so much big science as it is coordinated, interdisciplinary science. Especially in the sequencing part of the project, both the generation of the data and particularly their interpretation will require the recruitment of experts from disciplines that have had little involvement in biology to date. It seems clear that handling the information, validating, storing, and retrieving it, searching it for patterns indicative of functional domains, and identifying the coding portions will require much computer-assisted expertise. This represents a formidable challenge to the information scientist.

Regarding the argument that an institution where genomics (a generic term for mapping and sequencing) is being done is a poor site for graduate training and that the Human Genome Project will have a major adverse effect on graduate programs in biology, reality may be quite the contrary. With the full map and sequence, graduate students will be spared the drudgery of cloning and sequencing particular genes before they can get down to the much more interesting and intellectually demanding work of studying variation, function, regulation, and so on. The genomics laboratories will be superb settings for training a new breed of scientist—one who is prepared to capitalize on both the molecular genetics revolution and the computation revolution. These will be the leaders in biology in the twenty-first century.

It appears that the Human Genome Project is being made a scapegoat by those frustrated by the tight funding for research. If it was discontinued, the funding situation would not change perceptibly. Indeed, the excitement stimulated by discussion of the Human Genome Project has had and can contiue to have a beneficial effect on science funding generally.

The Societal Implications of the Human Genome Project

Attention to the ethical, legal, and social issues (ELSI) involved in the Human Genome Project has been unprecedentedly early and intensive—and appropriately so. The main concerns surround the use, misuse, and abuse of the information generated by the Human Genome Project and the power it provides for discerning the genetic makeup of individuals. Since this is the topic of this book, I will here make only some general statements.

It is my impression that the "problems" potentially created by the information of the Human Genome Project are not qualitatively different from those encountered every day in the practice of clinical genetics. They are, however, much greater in scope and significance. There are two generic risks. First, the information from the Human Genome Project and its ancillary endeavors will widen the gap between what we can diagnose and what we can do anything about (treat). This is already a problem in relation to Huntington disease and many other genetic disorders. A second hazard is that the gap will be widened between what we (physicians, scientists, and public alike) think we know and what we really know. In the years to come, weak associations will be found between particular genomic constitutions and specific behavioral peculiarities or susceptibilities to disease. Many of these, in the last analysis, will prove spurious. The significance of others, although statistically validated, will be blown out of all proportion, to the disadvantage of individuals and society as a whole.

The increasing gap between what we think we know and what we really know relates also to the hazards of reductionism and determinism.[23] Absurd as it may sound, the impression may become prevalent that when we know the sequence of the human genome down to the last nucleotide, we know all it means to be human—the ultimate of reductionism. Furthermore, it may become generally assumed that there is a direct one-to-one relationship between genomic constitution *and* all aspects of human health, disease, and behavior—the ultimate of determinism and hereditarianism.

NOTES

1. National Research Council, Committee on Mapping and Sequencing the Human Genome, *Mapping and Sequencing the Human Genome,* National Academy Press, Washington, D.C., 1988.
2. M. Olson, L. Hood, C. Cantor, and D. Botstein, "A Common Language for Physical Mapping of the Human Genome," *Science* 245:1434-5, 1989.
3. V. A. McKusick, *Mendelian Inheritance in Man: Catalogs of Autosomal Dominant, Autosomal Recessive, and X-Linked Phenotypes,* 9th ed. Johns Hopkins University Press, Baltimore, 1990.
4. E. B. Wilson, *The Cell in Development and Inheritance,* Monograph, 1896, pp. 182-185. V. A. McKusick, "The Morbid Anatomy of the Human Genome: A Review of Gene Mapping in Clinical Medicine," *Medicine* 65:1-33, 1986; 66:1-63, 237-96, 1987; 67:1-19, 1988.
5. E. B. Wilson, "The Sex Chromosomes," *Arch. Mikrosk. Anat. Enwicklungsmech,* 77:249-71, 1911.
6. J. F. Horner, "Die Erblichkeit des Daltonismus," *Amtl. Ber. ueber die Verwaltung des Medizinalwesens des Kt. Zurich v. Jahre 1876,* 31:208-11, 1876.
7. J.B.S. Haldane, A. D. Sprunt, and N. W. Haldane, "Reduplication in Mice. (Preliminary Communication)," *Journal of Genetics* 5:133-5, 1915.
8. J. Ott, *Analysis of Human Genetic Linkage.* Johns Hopkins University Press, Baltimore, 1985.
9. R. P. Donahue, W. B. Bias, J. H. Renwick, and V. A. McKusick, "Probable Assignment of the Duffy Blood Group Locus to Chromosome I in Man," *Proceedings of the National Academy of Sciences USA* 61:949-55, 1968.

10. D. Botstein, R. L. White, M. Skolnick, and R. W. Davis, "Construction of a Genetic Linkage Map in Man Using Restriction Length Polymorphisms," *American Journal of Human Genetics* 1980; 32:314–31, 1980.

11. R. G. Langlois, L-C. Yu, W. Gray, and A. V. Carrano, "Quantitative Karyotyping of Human Chromosome by Dual Beam Flow Cytometry," *Proceedings of the National Academy of Sciences USA* 79:7876–80, 1982.

12. L. Roberts, "The Worm Project," *Science* 248:1310–3, 1990.

13. Y.-s. Fan, L. M. Davis, and T. B. Shows, "Mapping Small DNA Sequences by Fluorescence In Situ Hybridization Directly on Banded Metaphase Chromosomes," *Proceedings of the National Academy of Sciences USA* 87:6223–27, 1990. J. B. Lawrence, C. A. Villnave, and R. H. Singer, "Sensitive High Resolution Chromatin and Chromosome Mapping In Situ: Presence and Orientation of Closely Integrated Copies of EBV in a Lymphoma Cell Line," *Cell* 52:51–6, 1988.

14. See note 10.

15. M. Boehnke, N. Arnheim, H. Li, and F. S. Collins, "Fine-Structure Genetic Mapping of Human Chromosomes Using the Polymerase Chain Reaction on Single Sperm: Experimental Design Considerations," *American Journal of Human Genetics* 45:21–32, 1989. X. Cui, H. Li, T. M. Goradia, K. Lange, H. H. Kazazian, Jr., D. Galas, and N. Arnheim, "Single-Sperm Typing: Determination of Genetic Distance between the G-Gamma-Globin and Parathyroid Hormone Loci by Using the Polymerase Chain Reaction and Allele-Specific Oligomers," *Proceedings of the National Academy of Sciences* 86:9389–93, 1989.

16. E. S. Lander, and D. Botstein, "Homozygosity Mapping: A Way to Map Human Recessive Traits with the DNA of Inbred Children," *Science* 235:1567–70, 1987.

17. Homozygosity mapping bears similarities in principle to mapping in mice by use of recombinant inbred strains. B. A. Taylor, "Recombinant Inbred Strains." In: M. Lyon and A. G. Searle (eds.), *Genetic Variants and Strains of the Laboratory Mouse,* 2nd ed., Oxford University Press, New York, 1989.

18. S. J. Goss, and H. Harris, "New Method for Mapping Genes in Human Chromosomes," *Nature* 255:680–4, 1975.

19. P. J. Goodfellow, S. Povey, H. A. Nevanlinna, and P. N. Goodfellow, "Generation of a Panel of Somatic Cell Hybrids Containing Unselected Fragments of Human Chromosome 10 by X-Ray Irradiation and Cell Fusion: Application to Isolating the MEN2A Region in Hybrid Cells," *Somatic Cell Molecular Genetics* 16:163–71, 1990. D. R. Cox, C. A. Pritchard, E. Uglum, D. Gashere, J. Kobori, and R. M. Meyers, "Segregation of the Huntington Disease Region of Human Chromosome 4 in a Somatic Cell Hybrid," *Genomics* 4:397–407, 1989.

20. See note 3 (9th ed., 1990).

21. V. A. McKusick, "The Human Genome Organization: History, Purposes, and Membership," *Genomics* 5:385–7, 1989.

22. W. Gilbert, "Towards a Pardigm Shift in Biology," *Nature* 349:99, 1991.

23. Discussed in detail by E. Shuster, Chapter 6, this volume.

II

Social Policy Implications

3

Those Who Forget Their History: Lessons from the Recent Past for the Human Genome Quest

Judith P. Swazey

The national and international quest to map and sequence the human genome is remarkable in a number of respects, in addition to the explosive advances in molecular biological knowledge and techniques that have made its inception possible. As various commentators have noted, there are two particularly striking and perhaps unique features of the Human Genome Project. One is the focused manner in which the potential applications of the knowledge the project may engender, much more so than the mapping and sequencing per se, raise virtually the whole panoply of ethical, legal, human value, and social policy issues that have been generated by basic and applied human genetics research.[1] Second, compared to our usual after-the-fact responses to the implications of other biomedical endeavors, many of those engaged in the project have realized from the outset "that serious social policy and ethical issues are raised by the research, and that steps ought to be taken now to try to assure that the benefits of the project are maximized and the potential dark side is minimized."[2]

Based on my own forays into social, ethical, legal, and policy issues raised by work in human genetics and other biomedical arenas,[3] my hope—however faint—is that two things will happen as the mapping and sequencing work and their applications go forward. First, I hope that many of the major players in the project—the scientists themselves—will exhibit more than a pro forma recognition of the implications and issues in which their work is embedded.

45

The "ELSI" components of the genome work being supported by the NIH and DOE provide an opportunity for the scientist-shapers of the project to affirm, "Look how ethical and socially responsible we are being." ELSI—an imagistically unfortunate acronym—certainly is being taken seriously by the social scientists, ethicists, lawyers, and assorted other scholars, who have seldom had such financial largesse available to them, and their studies should yield a body of interesting and in some case practically useful findings and recommendations. But in both the short term and the long run, the significance of the ELSI component will be greatly diminished if the concerns that generated it, and its work and results, are seen by scientists and clinicians as politically necessary but basically irrelevant appendages to the "real work" of the Genome Project. The far-reaching import of the Genome Project should be treated as an ethically as well as biomedically "serious" venture, one that demands and hopefully will receive reasoned, proactive thought, deliberation, and debate from broadly representative sectors of society, and, as needed, sociomedical policy actions.

If the Human Genome Project's ramifications are treated as "serious things," my second hope is that those engaged in dealing with the issues will draw on the accumulated body of experiences and knowledge from past excursions in basic and applied human genetics. One of our tendencies in American society is to rapidly forget, or ignore, even our recent history and all that it can teach us and thus to constantly and often painfully rediscover and reexperience patterns of issues and the concerns, actions, and reactions they generate.

The purpose of this chapter is to highlight a few important messages or lessons from the not-very-distant past, which can inform the workings of the Human Genome Project and the attempt to unravel and effectively address the myriad issues it raises. The four "lesson areas" involve (1) the "world view" undergirding the Human Genome Project, (2) the "mythos" or mythology about the project that its advocates are creating, (3) some particularly serious issues concerning the social power of biological information, and (4) some sociopolitical strategy lessons from the recombinant DNA research controversy.

The "Cultural Shaping" of the Human Genome Initiative

Science and technology, definitions of health, disease, and illness, and the kinds of medicine we practice are not entirely objective, rational, value-free, and culturally neutral. They are shaped, often powerfully, by the culture of a society and in turn help to shape the values, attitudes, and beliefs of the people living in a particular society during a given historical period.[4] Those involved in mapping and sequencing genomes should recognize, and others who will be affected by this work need to be made aware, that this endeavor is being shaped by a particular world view that has guided and motivated molecular biologists and other "makers of the Biological Revolution."

As analyzed by sociologist Howard Kaye, this "fundamental and deeply held" world view is a mechanistic and reductionistic one, which involves a "systematic attempt to reduce biology to the laws of physics and chemistry; organism to program; behavior to genes; life to reproduction; mind to matter; and culture to biology."[5] The conviction that science can and will succeed in understanding and precisely controlling nature and the human species through these successive reductionisms is reflected in the titles of some of the major writings of molecular biologists. Horace Judson, for example, titled his major study of the field *The Eighth Day of Creation,* referring to the biblical eighth day of redemption after the apocalypse. And one of molecular biology's pioneers, Gunther Stent, wrote *The Coming of the Golden Age: A View of the End of Progress,* in which he discussed how the discovery of DNA and of how to manipulate it had signaled an end to social and economic evolution.[6]

The Genome Project provides a fascinating example of how the reductionistic world view of molecular biology is fused with a secular utopianism that has been particularly characteristic of the American experience. As captured in the title of a book by historian Charles Rosenberg, *No Other Gods: On Science and American Thought,* we have deeply embedded sociocultural expectations about how the fruits of science and technology will enable us to create and dwell in a secular Garden of Eden.[7] In a blend of what seems to be genuine conviction and a zealous effort to market their costly project to funders, advocates of genome mapping have repeatedly voiced their expectations about the unparalleled opportunity this endeavor offers to conquer the "dragon of nature." "Way back" in 1986, for example, at an NIH advisory committee meeting convened to examine the embryonic project's potential and to provide guidance for future policy directions, one government official proclaimed that "this is the first time in 10,000 years of civilization that humans have had the capacity to upgrade their quality of life."[8] With even more deterministic hubris, James Watson has declared that the project provides "the ultimate tool for understanding ourselves at the molecular level." "We used to think," he has stated, that "our fate was in our stars. Now we know, in large measure, our fate is in our genes."[9]

What is particularly fascinating about the secular, mechanistic, and reductionistic world view of the Human Genome Project, and of molecular biology more generally, is the extent to which it is infused with religious language and imagery. Specifically, there are titles such as Judson's The *Eighth Day of Creation,* the famous "central dogma" of molecular biology, and references to the mapping effort as "the quest for the Holy Grail" that will create "the Book of Man."[10]

Whether wittingly or unwittingly, the makers of the molecular biological revolution seem to be trying to persuade us, and perhaps themselves, of science's God-like ability to master nature and the Great Chain of Being. To the extent that they are inclined to ponder their ideology, it would be useful for molecular biologists to unravel and understand the reasons for the language and imagery they have adopted and decide how well, in the end, it serves their cause and their convictions. Conversely, scholars and policy makers should "tune in" to the rich imagery of the language in which the purposes and objectives of endeavors like the Genome Project are expressed. Some scholars, like philosopher Ruth Macklin, dismiss the "Holy Grail" and "Book of Man" imagery as silly remarks that "no one takes seriously." But I take them seriously, because they convey important messages about what sociologists call the deep-structured meaning of molecular biology to members of its scientific community.

Genomes, Maps, and Myths

The omnipresence of myths in all societies and cultures, and in communities such as those formed by groups of scientists, technologists, and entrepreneurs, indicates that there are powerful reasons for their creation, transmission, and perpetuation. But depending in part on why they are created and how and why they are used, myths can be dysfunctional as well as functional, a double-edged property that I see in the mythos of the Human Genome Project. In a recent book about the workings and aspirations of the biotechnology industry, aptly titled *Gene Dreams,* Robert Teitelman writes that "despite its youth, the reality of biotechnology is often lost within a fog of its own enthralling mythology."[11] Since the even younger genome initiative employs the tools of the "new biotechnology," and since both spring from and partake of molecular biology and its world view and sociocultural values, one might expect that the Human Genome Project has been enshrouded in a similar mythology. From even a cursory review of articles, papers, and reports, this seems to be the case.

Taken together, many of the messages about the Project that have been conveyed to the scientific community, Congress, various funding sources, and the public weave together strands of secular utopian optimism and promises, the fervor of a religious crusade, and the "we shall overcome" ethos of the Western pioneers. The human genome is "the last frontier" of human anatomy to be conquered, and when the "Book of Man" has been charted and deciphered, we will have the "ultimate answers to the chemical underpinnings of human existence." Those answers, in turn, will help us understand and conquer a wide range of private and public health problems and will at last make it possible for "eugenic goals to be achieved through technological as opposed to social controls."

At this stage in the mapping and sequencing effort, it is hard to tell how much of the Genome Project's mythos depicts its progenitors' genuine convictions about what the effort will yield, and how much of it is a Madison Avenue type of hyperbole, put forth to defend the Project from critics. As many of the Project's champions have recognized, there are perils in reacting to concerns and criticisms about its content and goals, organization, and costs and financing by attempting to "oversell" it. Mapping and sequencing the genomes of humans and other species is an enormous task, and there is a real potential of creating a crisis of failed expectations about how much can be accomplished, how soon, and with what benefits to health and human welfare.

The Social Power of Biological Information

Perhaps the most sociopolitically, ethically, and emotionally potent set of issues contained within the orbit of the Genome Project involves the social power of biological information. That power, both real and perceived, and its uses for good and for ill have been a core issue in the unfolding history of efforts to prevent and treat disease and dysfunction through applied human genetics and the various efforts that have been made in the name of eugenics to reproductively contain or obliterate the "unfit" or to enhance and "improve" the "fit."[12]

Some strong caveats about the social power of biological information are proffered by Nelkin and Tancredi in their book *Dangerous Diagnostics.* They argue that the "new diagnostics"—tests that are being used or developed from research in genetics and the neurosciences—have great promise for understanding and more effectively treating many diseases, but also great potential for uses or misuses that could create a new "biological underclass." The increased preoccupation with and use of diagnostic testing, they argue, "reflects two cultural tendencies in American society: the actuarial mind-set, reflected in the prevailing approach to problems of potential risk, and the related tendency to reduce these problems to biological or medical terms."[13]

One aspect of this biological reductionism and determinism that most concerns analysts like Nelkin and Tancredi is the social power of information about our individual and once-private biology. They are alarmed, for example, by signals that diagnostic genetic testing is being used to support the social values and social policies of a new, more sophisticated version of eugenics. "For the most part," they note, "the new eugenics" that has surfaced over the past 20 years "has avoided generalizations about class and race, focusing instead on the individual benefits that follow from genetics research."[14] Particular attention, and controversy, has focused on individual rights and responsibilities concerning "reproductive fitness." Prominent geneticists, for example, have assumed, or urged the enactment of policies to require, that people will use "the . . . new biology to assure the quality of all babies. . . . No parent will have the right to burden society with a malformed or mentally incompetent child."[15] To foster this drive for genetically "healthy" or "fit" persons, there are organizations such as the Eugenics Special Interest Group, founded in 1982 as a part of MENSA. Its objectives are to "provide a communications network for all people committed to enhancing human genetic quality; to develop sound, innovative projects to produce eugenic benefit; . . . and to enlighten the public about the unique benefit of eugenics with the intention of ultimately influencing public policy."[16]

Moreover, although nominally concerned with the positive rather than the negative applications of the new genetic knowledge and techniques, disquieting references to the "pollution of the gene pool," "genetically healthy societies," and "optimal genetic strategies" have crept into the scientific literature of the 1980s and 1990s. Many people, too, have found a chilling negative neo-eugenicism implicit in suggestions that knowledge from the genome initiative will enable society to solve problems such as homelessness and poverty.[17]

Medical-geneticist Robert F. Murray, Jr. has argued that the concept of "genetic health" is highly uncertain and problematic and on close analysis is "erroneous," "probably meaningless," and, above all, "dangerous."[18] What has long concerned me and like-minded colleagues is the ways that some of the uses of genetics reflect, and reinforce, a value system that contains an intolerance for "imperfection." In this case, the concerns have to do with what we are defining as genetic perfection, or at least health and fitness, and the types of powerful social mechanisms we will allow to be used to monitor, sanction, or otherwise control those who deviate from those norms. While not overly given to Orwellian fears, I concur with Nelkin and Tancredi's observation and warning that

> [t]esting for the biological origins of disease can affect our concepts of social equality, justice, and privacy, and our ideas about choice and free will. . . . The

most important implication of biological testing is the risk of expanding the number of people who simply do not fit. . . . What is to be defined as normal or abnormal, able or disabled, healthy or diseased? And whose yardstick should prevail? . . . [E]ven if there was perfect predictive information, we should not underestimate the dangers of a new eugenics. If biological tests are used to conform people to rigid institutional norms, we risk reducing social tolerance for the variation in human experience. We risk increasing the number of people defined as unemployable, uneducable, or uninsurable. We risk creating a biologic underclass.[19]

Another dimension of the social power of biological information that also has surfaced in genetic screening and testing, and that will be greatly magnified by the fruits of the Genome Project, is the effect on individuals of knowledge about "anomalies" in their genetic makeup. Medical geneticists and counselors need to dwell long and hard on how they and their clients/patients should handle the psychosocial implications of the ability to detect genotypic variations that may, or may not, have clinical significance for the health of the individual or for her or his progeny. Knowledge that a person carries a "genetic susceptibility" to one or more multifactorially caused diseases, for example, can greatly magnify the sense of being at risk for various types of diseases and illnesses and swell the ranks of the "worried well" in a society that clings to the mirage of health described so eloquently by the late René Dubos.

Polity and Politics: Lessons from the rDNA Controversy

Participants in the Human Genome Project have been, and will be, enmeshed in at least two sociopolitical and values arenas that will test their skills at charting and navigating through turbulent and sometimes treacherous seas. The first arena, which I will only mention, involves resource allocation battles about the funding priorities and amounts being given to mapping and sequencing projects. In an era of fiscal stringency, those battles are likely to continue and probably intensify, given the "big science" aura that the congressionally approved fifteen-year Human Genome Project now bears. In this arena, it is almost irrelevant as to whether or not those involved in the Genome Project view their work as biology's entrance into "big science," and how small or large the money pot is relative to federally funded ventures in other fields of science and technology. What needs to be addressed in the political give-and-take of federally funded research are the perceptions and beliefs of knowledgeable critics about whether the Project warrants its special organizational and fiscal statuses. The mapping and sequencing work, these critics feel strongly, could and would go forward without the special priorities that have been bestowed on it by Congress "as the result of skillful persuasion [by] distinguished scientists."[20] "I like much of the science of the project and I admire the scientists involved," embryologist Donald Brown of the Carnegie Institution of Washington wrote in a letter to *Science.*

> [But] I dispute the "top down" mechanisms by which it is being administered. It is budgeted directly from Congress like a separate [NIH] institute, with its own administration and council and study sections. As a result, it is overbudgeted, overtargeted, overprioritized, overadministered, and micromanaged.
> . . . The project is supposed to be an add-on to the NIH budget, but even genome administrators concede that its funding inevitably will compete with the rest of the NIH budget. If this is so, then its science should be considered in competition with the rest of NIH science.[21]

The second arena deals with the ethical and human value concerns, legal issues, and social policy matters that may arise in foreseen or as yet unanticipated ways through the uses of knowledge about the human genome. For those who have been or will be drawn into the discussion and debate about the content of these issues, strategies for dealing with them, and how they should be resolved, the recent history of the recombinant DNA research controversy during the 1970s may provide some guidance about the governance of science as a social institution.

The rDNA controversy began when a small group of scientists went public with their concerns about the potential hazards of new techniques for forming rDNA molecules. The scientific issues they raised rapidly escalated into a social, political, legal, and philosophical, as well as scientific, cause célèbre, which was played out in a variety of local, state, federal, and international arenas.[22] Without going into the details of the content and dynamics of the rDNA controversy, let me briefly draw out six "lessons" for scientists engaged with the implications of mapping and sequencing the human genome.

Lesson 1. Acknowledge and address the implications of your work, but don't try to deal with them unilaterally. The scientists who took the unusual step of calling for and then adhering to a moratorium[23] on certain types of rDNA experiments until their potential biohazards could be assessed were exercising what they and many others viewed as a high degree of social responsibility.

Their actions were premised, in part, on an ethos or normative principle that is strongly held by scientists, physicians, academics, and other professionals in modern Western societies: by virtue of their training, expertise, and the nature of their work, scientists [and other professionals] are and should be treated as a self-regulating "company of equals."[24] Subsequently, however, these recombinant DNA researchers learned anew that science is inescapably a social institution, that its practitioners will be criticized as well as praised for such self-governance efforts, and that, however well intentioned and well executed, internal policy making may not ward off external regulation. From within their own ranks, for example, Erwin Chargaff, in the religious language so common to molecular biology, excoriated the "molecular bishops and church fathers" who came to the "Council of Asilomar" in February 1975 to develop recommendations for how rDNA research should proceed.[25] And in a post-Asilomar speech about the research, Senator Edward Kennedy at once commended the scientists for "attempt[ing] to think through the social consequences of their work" and chastised them for "making public policy in private."[26]

Lesson 2. Even though you may suffer a backlash from your efforts to exercise social responsibility and feel that "no good deed goes unpunished," don't disavow your intent or actions, and retreat to the haven of the laboratory. Science is a social institution, and scientists, as members of that amorphous entity called "the public," need to participate in public discourse about science and science policy. Those scientists who share their scientific expertise in public as well as professional forums provide a vital ingredient for an informed discourse. At the same time, however, when they venture into ethical, legal, and social policy domains, scientists should be on guard against the generalization of expertise: that is, the assumption that one's knowledge of a particular, specialized area qualifies a person to speak with equal authority on matters that may require quite different types of expertise.

Lesson 3. Don't try to completely disaggregate the scientific and social or ethical aspects of your work, and treat the latter as distant, tangential matters. David Baltimore did this in his opening statement at the 1975 Asilomar meeting, when he reminded the participants that they were there to focus on scientific issues rather than "peripheral" ethical and moral questions about the uses of rDNA technology.[27] While such a charge may have been appropriate to the organizing committee's agenda for Asilomar, it perpetuates the mythology of "value-free science" and fuels concerns that scientists just want to get on with doing their science, regardless of the consequences.

Lesson 4. "Here we are," a young scientist lamented during the Asilomar conference, "sitting in a chapel next to the ocean, huddled around a forbidden tree, trying to create some new commandments—and there's no goddamn Moses in sight."[28] As he and others discovered during the course of the rDNA controversy, the process of identifying and grappling with, much less deciding how to resolve, the issues posed by advances in scientific knowledge and technique are seldom, if ever, simple and amenable to a self-evident and rapid closure. This is true even within the confines of a scientific community, and as LeRoy Walters discusses in Chapter 16 with respect to the federal Recombinant Advisory Committee, the complexity and time frame of the process usually increase apace with the numbers of constituencies involved.

Lesson 5. Don't demean or discount the ability of laypersons to understand and deal with technically complex scientific matters. In Cambridge, Massachusetts, a lay Laboratory Experimentation Review Board was appointed to decide whether rDNA experiments could be done within the city's confines, and if so, under what conditions. To the astonishment of many scientists, the Board members proved they were eminently capable of understanding the relevant science and technology and of proposing conditions for P3 research that, many felt, were more sophisticated than those formulated by the scientists of the NIH Advisory Committee.[29]

The issues posed by potential uses of knowledge about the human genome hopefully will receive the thoughtful involvement of widely representative sectors of our society. In addition to the governmental commission model discussed by LeRoy Walters, another possible model, which could involve a broader grass-roots dialogue, is suggested by the various "community bioethics" groups at work on health care issues at local and state levels across the country.[30]

Lesson 6. Don't belittle or trivialize the layperson's "monster mythology"[31] fears about the fruits of science and technology. The rDNA controversy was replete with scenarios about the havoc that would be wreaked by "Frankenstein's little bugs" let loose on humans and nature. It is not sufficient, fair, or accurate to dismiss the fears of the public as trivial and unrealistic, as has been done by scientists like Sir Peter Medawar: "[F]or their excess of fearfulness, laymen have only themselves to blame and their nightmares are a judgment upon them for their deep-seated scientific illiteracy . . ."[32] Scientists, too, find meaning and uses in our culture's mythology, for scientific endeavors like rDNA research and the Human Genome Project "penetrate to and draw forth our deepest feelings about man's capacity to understand and manipulate the fundamental phenomena of life. They are feelings compounded of awe, promise, and unease, and they should not be dismissed lightly."[33]

NOTES

1. See, for example: G. J. Annas, "Mapping the Human Genome and the Meaning of Monster Mythology," *Emory Law Journal* 39:640, 1990. Capron does not

believe that the Human Genome Project "*in* and *of itself* poses any major ethical problems," but Capron holds that "some of the knowledge gained through mapping the genome may raise significant issues," particularly in the context of genetic screening (A. M. Capron, "Which Ills to Bear? Reevaluating the 'Threat' of Modern Genetics," *Emory Law Journal* 39:682, 1990).

2. Annas, 1990, p. 639.
3. See, for example: J. R. Sorenson, J. P. Swazey, and N. Scotch, *Reproductive Past, Reproductive Futures: Genetic Counseling and Its Effectiveness.* Birth Defects: Original Articles Series, Vol. XVII, No. 4; J. P. Swazey, "Phenylketonuria, A Case Study in Biomedical Legislation," *Urban Law Review* 48:883–931, 1971; J. P. Swazey, J. R. Sorenson, and C. B. Wong, "Risks and Benefits, Rights, and Responsibilities: A History of the Recombinant DNA Research Controversy," *Southern California Law Review* 51:1019–78, 1978.
4. A number of illustrative case studies are provided in a special issue of the *International Journal of Technology Assessment in Health Care* (Vol. 2, 1986), edited by Renée C. Fox, on "The Cultural Shaping of Biomedical Science and Technology."
5. H. Kaye, "The Biological Revolution and Its Cultural Context. In Fox, 1986, p. 278.
6. H. Judson, *The Eighth Day of Creation.* Simon & Schuster, New York, 1979; G. Stent, *The Coming of the Golden Age: A View to the End of Progress.* University of California Press, San Francisco, 1969.
7. For studies of secular utopianism, especially of the technological variety, in the history of American society and its culture, see: L. Marx, *The Machine in the Garden,* Oxford University Press, London, 1964; H. Segal, *Technological Utopianism in American Culture,* University of Chicago Press, Chicago, 1985.
8. Summary of discussion by D. B. Merrifield, in *The Human Genome.* Proceedings of the 54th Meeting of the Advisory Committee to the Director of the National Institutes of Health, Oct. 16–7, 1986, p. 18.
9. Quoted in L. Jaroff, "The Gene Hunt," *Time,* March 20, 1989, pp. 63, 67.
10. See, for example: Hall, "James Watson and the Search for Biology's 'Holy Grail'," *Smithsonian,* February 1990, p. 47.
11. R. Teitelman, *Gene Dreams. Wall Street, Academia, and the Rise of Biotechnology,* Basic Books, New York, 1989, p. 4.
12. For some thought-provoking analyses of eugenics, see: D. J. Kevles, *In the Name of Eugenics.* University of California Press, Berkeley, 1985; R. J. Lifton, *The Nazi Doctors,* Basic Books, New York, 1986; B. Muller-Hill, *Murderous Science,* Oxford University Press, New York, 1988; R. Proctor, *Racial Hygiene,* Harvard University Press, Cambridge, MA, 1988.
13. D. Nelkin and L. Tancredi, *Dangerous Diagnostics. The Social Power of Biological Information,* Basic Books, New York, 1989, p. 9.
14. Ibid., p. 13.
15. B. Glass, "Science: Endless Horizon or Golden Age?" *Science* 171:23–29, 1971. For recent analyses of some of the ethical and legal issues in the junction between genetics and reproduction, see: J. Robertson, "Procreative Liberty and Human Genetics," *Emory Law Journal* 39:697, 1990; S. Elias and G. Annas, *Reproductive Genetics and the Law,* Year Book Medical Publishers, Chicago, 1987.
16. Nelkin and Tancredi, 1989, p. 13.

17. See, for example, two editorials by *Science* editor D. Koshland: "Sequences and Consequences of the Human Genome," *Science* 246:189, Oct. 13, 1989; and "The Rational Approach to the Irrational," *Science* 250:189, Oct. 12, 1990.

18. R. Murray, "Genetic Health: A Dangerous, Probably Erroneous, and Perhaps Meaningless Concept." In *Genetic Counseling: Facts, Values, and Norms,* edited by A. Capron, M. Lappe, R. Murray, et al. Birth Defects: Original Articles Series, Vol. IX, No. 2, Alan R. Liss, New York, 1979, Ch. 5.

19. Ibid., pp. 167, 175–6.

20. D. D. Brown, [Letter to the editor]. *Science* 251:855, Feb. 22, 1991.

21. Ibid., p. 854.

22. For a narrative history and analysis of the rDNA controversy from 1972 to 1978, see Swazey, Sorenson, and Wong, 1978.

23. To my knowledge, the calling of a moratorium in basic research is a very infrequent phenomenon, in contrast to the more common occurrence of moratoria in the early phase of clinical research. For case studies of moratoria in clinical research and an analysis of the factors that can impel or deter a moratorium, see R. Fox and J. Swazey, *The Courage to Fail,* 2d ed., rev., University of Chicago Press, Chicago, 1978, Ch. 5; and J. Swazey, R. Fox, and J. Watkins, "Assessing the Artificial Heart Experiment: The Clinical Moratorium Revisited," *International Journal of Technology Assessment in Health Care* 2(3):387–410, 1990.

24. B. Barber, *Science and the Social Order,* Free Press, Glencoe, IL, 1952, pp. 139–56.

25. Swazey, Sorenson, and Wong, 1978, pp. 1034–5.

26. Ibid., p. 1037 N.45.

27. Ibid., p. 1031.

28. Ibid., p. 1015.

29. Ibid., pp. 1057–63.

30. B. Jennings, "A Grassroots Movement in Bioethics. Special Supplement," *Hastings Center Report,* June/July 1988.

31. Annas, 1990.

32. Swazey, Sorenson, and Wong, 1978, p. 1076.

33. Ibid., p. 1077.

4

Genomics and Eugenics: How Fair Is the Comparison?

Robert N. Proctor

In October 1990 the U.S. Department of Energy and the National Institutes of Health formally launched the Human Genome Project, the goal of which is to map and sequence the human genetic material—all of the estimated 100,000 human genes and three billion nucleotides that make up the six-foot-long DNA molecule. The project is supposed to take fifteen years and to cost some $3 billion, making it the most ambitious undertaking in the history of biology. Comparisons have been drawn to the Apollo moon launch or the Manhattan Project, though the project is as often described in biblical or epic terms. Former project leader James Watson has compared the mapping of the genome to the mapping of the American western frontier; Mark Pearson, Du Pont's director of molecular biology, has stated that the project will usher in "the Golden Age of molecular medicine."[1] A recent joint DOE and Health and Human Services report heralds the project as generating information that will constitute "the source book for biomedical science in the 21st century."[2] Popular accounts by journalists border on the giddy, asking us to think of the project as "the most astonishing scientific adventure of our time," and so forth.[3]

Why the effort to map the genes? It is hard to believe, with Robert Sinsheimer, that the genome deserves sequencing simply "because it is there."[4] A lot of things are out there and we don't pour billions into researching them. True, there are fascinating problems of a relatively "pure science" nature that may be solved from the knowledge and techniques gained from the project.

57

Some of these are of historical interest. In February 1991, for example, the National Museum of Health and Medicine announced it was considering proposals to map Abraham Lincoln's genes, using samples of hair, blood, and other tissues preserved from his assassination. Genetic analysis might reveal whether he suffered from Marfan syndrome, a genetic disease that makes one tall and gangly and shortens one's life.[5] Paleontology is also likely to get a boost from the new techniques. In March 1991, scientists at Stanford announced the recovery of genetic material from the brain cells of a man preserved in a central Florida peat bog for more than 7,500 years; efforts are now underway to sequence the material. As recovery techniques improve, DNA is likely to be recovered from older and older organisms preserved in ice, ash, or tar pits. Similar techniques (using protein sequencing, for example) were responsible for showing that apes and humans share a common ancestor as recently as five or six million years ago; sequence data have helped to confirm Darwin's hypothesis that modern humans evolved in Africa.[6] Detailed knowledge of our nucleotides, and especially of patterns of genetic variance, will no doubt unlock other secrets of our origins.[7]

Fascinating as such problems are, they are unlikely to have been the stimulus for a research effort of such magnitude. The Human Genome Project is supposed to be a project in pure science but, like most big science projects, there are also expectations of practical returns. One of the primary justifications for the project has been that it will provide the skills and knowledge required to begin treating genetic disease. In the past decade, genes have been identified for Huntington disease (on the tip of chromosome 4); Duchenne muscular dystrophy (X chromosome); cystic fibrosis (chromosome 7); colon cancer (chromosome 5); and 200 other diseases.[8] Hopes are high that tests will soon be found for many of the more common diseases from which millions suffer: in December 1990, Knight-Ridder newspapers carried front-page stories on the new genetics, listing some 90 million persons in the United States for whom genetic tests might be appropriate in the near future, including 500,000 carriers of adult polycystic kidney disease, 100,000 persons with fragile X syndrome, 65,000 persons with sickle cell anemia, 30,000 with cystic fibrosis, and 125,000 at risk for Huntington disease. The feature also reported that there are millions of others for whom there may be additional genetic risks, including 58,000,000 persons with hypertension, 15,000,000 dyslexics, 6,700,000 sufferers of atherosclerosis, five million persons with cancer, two million with manic-depressive illness, 1.5 million with schizophrenia, and so forth.[9]

It is hard to argue against a research program that promises at the least the possibility of treatments for these intractable diseases.[10] Still, many different ethical worries have been voiced concerning the project. Some have warned of a new eugenics; others worry we are in danger of destroying species integrity. The social use of the resulting knowledge is probably the most persistent source of uneasiness: a statement issued by the NIH working group coordinating research on the ethical, legal, and social implications of the Project (ELSI) warns that "if misinterpreted or misused, these new tools could open doors to psychological anguish, stigmatization, and discrimination" for people who carry diseased genes.[11]

There is much talk these days about the "social power of biological information" and the potential for abuse such power entails.[12] The argument I would like to make, by contrast, is that there is an equal danger from the abuse of genetic *misinformation.* Much of the historical abuse of genetics comes from an exaggerated belief that "nature" is more important than "nurture" in the expression of human talents and disabilities. "Genetic" is interpreted to mean "inevitable," and the role of socialization, diet, or some other aspect of the environment vital for genetic expression is ignored. One or another of these myths usually lies at the heart of most abuses of genetics. If the force of genes has been exaggerated, then there is as much a need to emphasize the social *impotence* of genetic information as its power. In an age where "It's genetic" has become a common rationalization for personal suffering and social inequality, this is particularly needed. Civil libertarians rightly worry about the prospects of social control that have emerged from the accumulation of genetic information, but I suggest that equal dangers will flow from exaggerations of the extent to which human behavior is genetic. These are concerns for which there is much historical precedent.

Eugenics

The social power of genetic information was not something discovered in the 1980s. Eugenics was the term coined in 1883 by Francis Galton (Charles Darwin's cousin and the inventor of identification by fingerprints) to designate "the study of the agencies under social control that may improve or impair the racial qualities of future generations, either physically or mentally."[13] Eugenics was more than a field of study, however. Eugenics became a popular political movement, providing counsel to governments on questions ranging from immigration and abortion to penal reform and psychiatric asylum. Eugenics was supposed to be long-run "preventive medicine," designed to eradicate human genetic illness before it could spread. Especially in Germany, racial hygiene (the German equivalent of Anglo-Saxon eugenics)[14] was considered one of the three prongs of responsible medical care: *personal* hygiene was the sphere of traditional medicine, focusing on the individual;

social hygiene was the sphere of public health, focusing on occupational safety and housing, clean air and water; and *racial* hygiene was the sphere of care for human germinal health, focusing on future generations. Racial hygiene was preventive rather than curative medicine; racial hygiene would provide tools for the long-run management of the human germplasm.[15]

Implicit in the work of most eugenicists was a set of fears: that "racial poisons" (especially alcohol, tobacco, narcotics, and syphilis) were threatening the health of the race; that the criminal, mentally ill, and morally dissolute were outbreeding the more upstanding elements of society. Eugenicists feared that the comforts of human civilization—notably welfare and medical care for the weak—had begun to erode the competitive struggle that normally maintains the fitness of animal populations; eugenicists worried that human compassion was allowing "subnormals" to survive and reproduce who otherwise, in a state of nature, would never have lived to bear children. Eugenicists believed that strong biological measures were needed to remove these inferiors from the pool of potential breeders and to encourage "the fit" to breed their kind. "Negative eugenics" would eliminate the weak (by sterilization, for example); "positive eugenics" would promote the healthy (by financial incentives for fitter families, for example). "Preventive eugenics" would safeguard the human genetic material by minimizing exposures to mutagens such as x-rays, lead, and alcohol.[16] State-sponsored management of the human germplasm was not, in this view, an unreasonable proposition given that, as H. H. Laughlin once wrote, the human germplasm belongs to society and not just "to the individual who carries it."

One of the pillars of eugenics ideology was biological determinism—the idea that biology lies at the root of most human talents and disabilities. Especially after 1900, with the rediscovery of Mendel's laws, eugenicists assumed that it was to genetics that we must look for the causes of crime, mental illness, and social deviance. Eugenicists tended to exaggerate the extent to which human behavior, human disease, and human institutions are biologically rooted. Traits as diverse as hernias, wanderlust, and divorce were assumed to be genetic. Eugenicists also believed that human populations vary widely in their susceptibility to disease. Otmar von Verschuer—Joseph Mengele's dissertation advisor—identified more than fifty genetic diseases in his 1937 treatise on "Genetic Pathology."[17] The Nazi government funded substantial research in this area: geneticists under the Third Reich sought to prove that alcohol addiction and schizophrenia were heritable traits, as were susceptibilities to infectious agents such as yellow fever and malaria (lower among blacks), dust-induced diseases such as black lung, and many physical deformities. Anthropologists sought to prove that Jews were more likely to suffer from physical deformities such as flat feet, while Germans were more likely to suffer from "urban" infections such as tuberculosis. (Jews had supposedly become immune to tuberculosis after centuries of urban living.) Professional journals devoted to "genetic health" and to "genetic pathology" were established—the first such journals in the world.[18]

The practical outcome of American eugenics and German racial hygiene is well known. In 1907, the state of Indiana passed the nation's first sterilization law, allowing the forcible sterilization of individuals suffering from any of a number of genetic defects. Over the next thirty years some thirty other U.S. states passed similar legislation, resulting in about 50,000 sterilizations by the end of World War II. Eugenicists also mounted efforts to bar immigration from undesirable racial stocks: the notorious immigration restriction act of 1924, passed after a number of eugenicists testified before Congress, reduced the flow of immigrants into this country from about 435,000 per year to less than 25,000 per year—a 95 percent reduction. The 1920s also saw new laws banning interracial marriage. In 1929, California banned marriage between whites and "Negroes, Mongolians, Mulattoes, or members of the Malay race." In 1926, Indiana's state legislature declared null and void marriages between whites and "persons having one-eighth or more of negro blood." By 1940 thirty American states had passed legislation barring racial miscegenation in one form or another; most of these laws were not repealed until after World War II.

German eugenics did not differ profoundly from American, except for the fact that it was far more thorough and enjoyed the support of an enthusiastic fascist government. The 1933 sterilization law ("Law for the Protection of Genetically Diseased Offspring") resulted in some 350,000 forced sterilizations by the end of the Nazi period; racial miscegenation laws barring marriage between Jews and non-Jews were comparable in effect to American laws. In certain respects the German laws were actually more lenient: in several U.S. states, for example, one part in thirty-two of Negro ancestry was sufficient to brand one as "colored"; in Germany, by contrast, persons with as much as one-fourth Jewish ancestry were for most practical purposes considered "German."[19] Genetic counseling became a prerequisite for obtaining a marriage license; laws were passed barring marriage between the genetically fit and the genetically unfit.

The relevance of eugenics for the present is severalfold. At one level, it presents a dramatic case of how genetic knowledge (and genetic ignorance!) can be coupled with repressive state policy to deprive individuals of rights and liberties. It also illustrates how scientists may lend their support to political movements, giving them an air of respectable legitimacy. Science served as a vehicle for the carriage and transport of popular prejudices. Geneticists were hardly in the vanguard of opposition to eugenics—indeed geneticists were among its most ardent supporters. In Germany, genetics was Nazi science par excellence: geneticists so rarely opposed the Nazi regime that party officials were astonished to find that Hans Stubbe, a prominent geneticist, had once been a member of the Communist Party (he later had applied to join the SS).[20] Medicine and biology were two of the most Nazified professions in Germany: half of all physicians joined the Nazi Party, along with about 60 percent of all biologists. Human geneticists were probably the worst in this regard: among ten full professors of anthropology in Nazi Germany (most of whom were human geneticists), eight were party members.[21]

When the Nazi regime collapsed in the spring of 1945, most prominent eugenicists were never punished. Most were reappointed to prestigious posts in the newly christened field of "human genetics"; the German postwar human genetics community grew directly out of the Nazi-era profession of racial hygiene.[22] The Nazi support for genetics (and vice versa) was largely forgotten in the postwar buildup; geneticists were understandably reluctant to investigate the complicities of their colleagues. As recently as 1987, Benno Müller-Hill, the German geneticist who codeciphered the lac-operon, could protest that "the rise of genetics is characterized by a gigantic process of repression of its history."[23]

The rise and fall of National Socialism did much to discredit the eugenics movement. American eugenicists had generally supported their German colleagues, but the atrocities of the war years ended popular support for the movement. Eugenics institutions were dismantled in Germany; eugenics legislation in the United States fell victim to a number of legal challenges. Concerns for the genetic health of the race continued, but only after efforts to break apparent links to the earlier movements. The British *Annals of Eugenics* was renamed the *Annals of Human Genetics* in 1954; the *Eugenics Quarterly* became the *Journal of Social Biology* in 1969. Postwar eugenicists worried about runaway population growth in India and Latin America, but also about the mutagenic effects of atomic radiation. Less than two months after Hiroshima, Guy I. Burch of the Population Reference Bureau asserted that the power of the atom had made it "more necessary than ever to pay attention to eugenics."[24] H. J. Muller, recipient of the 1946 Nobel Prize for his demonstration of x-ray-induced mutagenesis, warned that the atomic bombs over Hiroshima and Nagasaki had planted "hundreds of thousands of minute time-bombs in the survivors' germ cells"; Muller predicted that future descendants of those survivors might be so wracked by deformities they might well wish the bomb had killed them.[25]

Postwar government agencies were also eager to explore the health effects of atomic radiation. Geneticists were employed to explore the biological consequences of radiation-induced mutagenesis (James Neel by the U.S. Atomic

Bomb Casualty Commission; Otmar von Verschuer by West German officials); military officials in both countries launched elaborate research programs into how exposures at the workplace or bomb site might produce birth
defects, cancer, or other biological problems.[26] When the second edition of
the *Encyclopedia of the Social Sciences* was published in 1968, the article on
"Eugenics" identified the mutagenic effects of radiation as one of the central
concerns of a revitalized eugenics. The author, Gorden Allen, cautioned that
the mutations induced from the atomic bursts over Hiroshima and Nagasaki
were likely to exact a toll "in death and suffering spread over many generations."[27]

Department of Energy Interest

This brings us to the Human Genome Project. The U.S. government's drive
to map the human genome is a joint effort of the National Institutes of Health
(NIH) and the Department of Energy (DOE). NIH support for the project is
understandable given its position as the leading source of funds for U.S. biomedical research, but why is an agency responsible for the development and
regulation of energy technologies involved? DOE officials were among the
first to suggest the need for a large-scale sequencing effort, and by 1987 the
DOE was spending $4.2 million on human genomics. The NIH has now
eclipsed the DOE as the primary funder of the project (the DOE presently
provides about one-third of the funding for the project, the NIH the other
two-thirds), but it was in DOE laboratories that the idea for the project first
arose. Why were America's nuclear weapons laboratories spearheading the
effort to map the human genes?

The DOE (formerly the Atomic Energy Commission) has a long-standing
interest in human genetics, dating back to its early monitoring of the health
effects of the atomic bomb. Concerned about the effects of nuclear radiation
on human populations, the Atomic Energy Commission became one of the
major funders of genetics in the early postwar period. DOE scientists have
long been hampered, though, by their inability to measure precisely the
impact of radiation on human health. The Human Genome Project grew at
least partly out of DOE efforts to monitor radiation damage. In December
1984, scientists from DOE weapons laboratories meeting at the ski resort of
Alta, Utah, suggested that a massive sequencing effort might help identify the
kinds of genetic damage suffered by survivors of the Hiroshima bomb.[28] The
consensus of the meeting was that the mutagenic effects of atomic radiation
could not be adequately measured without substantial improvements in techniques of manipulating the human genome. James Neel, director of the

Atomic Bomb Casualty Commissions's mutation study, estimated that the mutation rate caused by the bombs dropped at Hiroshima and Nagasaki might be 10^{-8} per base per generation (or thirty new mutations per genome per generation), but to obtain an accurate figure one needed information on fifty billion base pairs from the children and four or five times this number from the parents. Many of those who participated in the Alta meeting, sponsored jointly by the DOE and the International Commission for Protection against Environmental Mutagens and Carcinogens, went on to develop sequencing strategies and techniques that would be used in the Human Genome Project.[29] By 1988 the DOE's Los Alamos Laboratory possessed the largest genetic data bank in the world—the GenBank—occupying an increasing fraction of the lab's 70 "Cray equivalents" of computer memory. In 1988, the GenBank already had an annual budget of $3.5 million, more than all other U.S. genetic and protein databases combined.[30]

Understanding radiation damage may have been one reason for DOE interest in the project, but cynics have suggested other factors. The desire to "beat the Japanese" surely played a role in obtaining initial funding for the project: according to Charles DeLisi, director of the DOE's Office of Health and Environmental Research, the Japanese had initiated a similar project some five years earlier than the United States, in 1980–1981. The Japanese project began at the RIKEN Institute in Tsukuba Science City, where efforts had been made to build an atomic bomb during World War II. The Japanese "lead" in automated sequencing technologies was later cited in Congress as a challenge for American policy makers to overcome.[31]

Still others have suggested that DOE involvement can be traced to the fact that the weapons labs were looking for ways to turn the enormous resources of the facilities toward civilian ends should the Reagan-era enthusiasm for military research ever begin to slacken.[32] Pete Domenici, U.S. Senator from New Mexico, made this clear at a May 2, 1987, meeting on the future of funding for the government's national laboratories, two of the largest of which—Sandia and Los Alamos, employing about 15,000 people—are in his home state. Arguing in favor of DOE involvement in the genome project, Domenici noted that the expertise of the labs could be turned from nuclear toward biological issues; by his own admission, he was worried about what might happen to the labs "if peace breaks out." David Botstein of MIT, an early critic of DOE involvement, was more blunt, describing DOE support for the project as "DOE's program for unemployed bomb-makers."[33]

The DOE's handbook describing the mission of its Human Genome Program states that a major goal of the DOE's health effects program is to develop "capacities to diagnose individual susceptibility to genome damage imposed by energy-related factors."[34] In January 1990, at the first workshop of the

Working Group on the Ethical, Legal and Social Issues Related to Mapping and Sequencing the Human Genome, I asked Benjamin Barnhart, manager of the DOE's Human Genome Program, why the department was so interested in genetic sequencing. He explained to me the history of DOE interest, but also noted in passing that the DOE was especially interested in those regions of the human DNA (chromosome 17, for example) where radiation repair genes are suspected to lie.

DOE interest in finding out who might have faulty radiation repair genes leads one to wonder whether knowledge gained from the project might be used for screening workers; critics worry that genomics may be used as yet another way to adapt workers to the workplace rather than vice versa.

Specters, Spurious and Genuine

Many criticisms have been raised against the Human Genome Project. Some have objected to its cost and its scale, fearing that the project will draw resources away from smaller-scale, equally worthy projects. Others have pointed out that sequencing is a tedious activity, and that it may be hard to motivate graduate students to work on the project. (Automation techniques are likely to relieve much of this tedium—indeed the relief of tedium through automation has been one of DOE's primary rationales for the project.) Still others have expressed concerns that the resultant knowledge is not likely to be of great value. James Walsh and Jon Marks, in a 1986 letter to *Nature,* suggested that sequencing the genome "would be about as useful as translating the complete works of Shakespeare into cuneiform." Bernard Davis was more diplomatic, suggesting that the goal of a complete and exhaustive map would require a lot of investigation into "junk" DNA. Surely, he argued, it would be better for mapping to proceed first in those areas of the chromosomes known to be of biological interest.[35]

By far the most persistent criticisms, however, have been those directed against the potential use of the knowledge once gained. In a project of such magnitude and intimacy (just how "intimate" people may regard knowledge of their genes is a matter one could certainly dispute), it is not surprising that ethical concerns have been raised. Several of these are genuine, some spurious.

Among the latter, spurious, concerns I would single out two: "playing God" and "perfect babies." Both are specters commonly raised when the presumption is that fundamental limits of common decency have been transgressed. In one common view, genetic engineers are likened to sorcerer's apprentices, assuming powers for which they have no right. "Meddling," "tinkering," and "forbidden fruit" are the common currency of this critique. Jeremy Rifkin champions this line of argument, along with the notion that once one begins to meddle there is a "slippery slope" along which one must fall: first we fix Huntington or cystic fibrosis, then less than average height or athletic ability, then left-handedness, and who knows what else?[36]

The problem with the abstract warning against "playing God" is that doctors presumably play God every time they treat an illness. All of human art or *techne* involves manipulating the natural environment; most medical technologies (including drugs) involve the manipulation of the human body. Abstract warnings against "playing God" may shock or slow research, but they do little to tell us what are the concrete dangers to be avoided, what policies to prefer.

A related and equally spurious specter appears in warnings against the quest for "perfect babies." Those who worry about discrimination or social stigma often fear that genomics is bringing about a quest for perfection that will obligate us to homogenize our children's genes the way we twist their teeth through orthodontics. Arguably, perfection is no more the goal of genetic than of any other kind of therapy. Noone seems to mind when parents provide their children with balanced diets or stimulating day care; why isn't the specter of perfection an issue here?

Genetic manipulation seems to call up additional fears. Among conservatives, what seems to bother people is that genetic technologies might allow people to make frivolous decisions concerning what kind of children they want to raise. *Time* magazine, reviewing the ethical perils of "treading on heredity," suggests that a woman might choose to abort a fetus for some minor genetic defect: "hardened" prochoice advocates are suspected of arguing that parents should have the right to abort "fetus after fetus until they get the 'perfect' baby."[37] *Time* magazine here conjures up an antiabortionist's fantasy that no prochoice advocate, so far as I'm aware, has ever suggested. Fears of abortion are (unfortunately) deeply entangled with the concerns about genetic testing. "Prolife" activists generally oppose all prenatal genetic testing on the grounds that abortion is one of the few practical alternatives to genetic disease.[38] The specter of women seeking "perfect babies" is often raised by antiabortionists worried about the emergence of yet another rationale for women to terminate a pregnancy. The American Society of Human Genetics confirmed these fears in May 1991 when it voted to recommend that antiabortion legislation include provisions for allowing termination of pregnancies where the fetus "is likely to have a serious genetic or congenital disorder."[39] Still, it is not an easy matter for a woman to go through months of pregnancy only to have it terminated; women are not likely to use abortions to fine-tune pregnancies on a lark.

The rhetoric of perfection also blends with the rhetoric of the "slippery slope." Those (Rifkin, for example) who warn that treating genetic defects will somehow lead to eliminating social undesirables also worry that the culturally "abnormal" will be stigmatized. But surely there are diseases the treatment of which will in no way imply invidious racial or cultural discrimination. As Nancy Wexler points out, suffering from Huntington disease is hardly something that crowns the glory of human diversity: "How is our humanity enhanced by permitting the perpetuation of diseases that devastate body and mind?"[40] Social stigma is certainly a danger (see below), but it is wrong to denounce as discriminatory or "playing God" every intervention in the human body. Rifkin and others who warn of a "new eugenics" should not forget that a blanket "no" to genetic manipulation is of little use in distinguishing lesser from greater evils. The constructive political task is to determine how safeguards can ensure that genetic manipulation promotes, rather than limits, human liberties. It is as easy to fall on the slippery slope of criticism as it is to slip on the slopes of technological optimism.

If perfection and playing God are spurious dangers, what are the genuine dangers? And what are we to make of the comparison between eugenics and genomics? The following six areas are ones for which concerns have been most commonly expressed.

Diagnosis without Therapy

A number of observers have noted that, in the field of genetic medicine, it is far easier to diagnose a disease than to recommend an effective therapy. Thousands of genetic diseases have been identified, but effective therapies are available for only a handful. Complicating the picture is the fact that many of these can be diagnosed years or even decades in advance of the actual onset of disease—often in utero, since the amniotic fluid surrounding the developing fetus (which one samples during amniocentesis) already contains cells with the full complement of fetal DNA.

Phenylketonuria (PKU) is one of the few examples of a genetic disease for which there is a successful therapy. Infants found to have the disorder can be placed on a restrictive diet to prevent mental retardation. But PKU is a relatively rare disorder. For the overwhelming majority of genetic diseases there is no effective treatment. Sickle cell anemia, the gene for which was identified in the 1960s, still has no effective cure. The same is true for Huntington disease, polyposis (a precursor of colon cancers), adult polycystic kidney disease (a progressive kidney disorder), and most of the various cancers for which a genetic predisposition has been found. For these, as for chromosomal anomalies like trisomy 21 (Down syndrome), prevention—through prenatal diagnosis and abortion—remains the only "therapy."

The time may come when treatment for many of these disorders will be made available, and a diagnosis of Alzheimer or Huntington disease will be no more serious than a diagnosis of PKU. But who can say how long this may take—twenty, forty years? Prior to the discovery of an effective treatment or cure (most likely in the form of genetic therapy), persons diagnosed with "genetic lesions" such as Huntington disease or cystic fibrosis will have to live with the knowledge that a future of debilitating illness awaits them. Never before have physicians been able to predict years or even decades in advance that specific individuals will contract a particular disease; the prospect has been described as "terrifying."[41] In such circumstances, many people may be unwilling even to undergo testing: among the 125,000 Americans at risk for Huntington's, for example, only 200 have come forward to be tested.

Some of the more troubling ethical implications of genomics stem from this fact of diagnosis without prospect of therapy. As in most ethical dilemmas, it is an artifact of social structure that ultimately generates the dilemma. The question of insurability is a case in point. If insurance companies come to regard genetic lesions as "preexisting conditions," it may be difficult for persons carrying those genes to obtain insurance. Physicians have already begun to face dilemmas of the sort described by Ray White, where patients found to be carriers of a gene predisposing them to colon cancer were not informed of their condition because this would prejudice their ability to obtain insurance (see also below).[42]

Some have predicted dramatically increased demands for genetic counseling, but until further therapies are available and costs for testing diminish, genetic counseling may not see the expansion many predict. People found to have genetic lesions will be able to choose how and whether to have children, but many will also have to live with the knowledge (if they choose to obtain it) that illness awaits them and that medicine may have little to offer.

Economic Injustice

Historical experience shows that new technologies work as often to exaggerate as to lessen economic inequalities. The agricultural Green Revolution ushered in a series of new, higher-yielding varieties of rice and wheat, but those who profited most from the new varieties were the wealthier farmers who could afford the pesticides and fertilizers required to grow them. Biotechnology can make new vaccines, but most of the diseases being investigated are First World diseases for which vaccines are often already available, although the real need is for vaccines against the scourges of the Third World—for which there is little economic incentive. In the mid-1980s, when Merck, Sharp & Dohme launched the world's first genetically engineered vaccine against hepatitis B, critics pointed out that most of those who need it (the poorer peoples of Asia) will never be able to afford it.[43]

Commercial exploitation generally requires patent rights. In 1980, the U.S. Patent Office awarded Ananda Chakrabarty of General Electric a patent for his oil-eating bacterium—the first biosynthetic microorganism ever to be patented. In 1988, Walter Gilbert of Harvard obtained a patent for a mouse he had engineered to develop breast cancer; the mouse, designed to be used by laboratories seeking therapies for the disease, was the first mammal (and indeed the first nonplant eukaryote) to receive a patent. E. I. du Pont now markets the Harvard mouse under the brand name "Oncomouse" to cancer researchers for about $50 per animal.[44]

The human genome is likely also to be commercialized soon; parts have already begun to be parceled up for sale. In 1987, Walter Gilbert of Harvard established the (short-lived) Genome Corporation to market databases containing sequences of key segments of the human genome. For nearly a decade the U.S. Patent and Trademark Office has been quietly granting patents for sequences of the human genome, including DNA sequences for plasminogen activator protein (1983), erythropoietin (1987), human T-cell antigen receptor (1987), and a number of others.[45] In the fall of 1991, scientific tempers flared when it was disclosed that NIH had applied for patents on hundreds of separate sequences of the human genome without even waiting to decipher the messages encoded by those sequences. Critics expressed fear that such action could incite the scientific equivalent of a land rush, with government and biotechnology companies racing to claim new genetic territory.[46]

New markets are also opening up in the area of genetic services. Companies with names like Genomyx, Genetic Therapy, Inc., and GeneScreen have been founded to develop genetic tests or therapies;[47] Focus Technologies is a biotechnology company developing genetic tests to be used by employers to screen workers. Demand for services has grown so rapidly in the area of forensic medicine that the American Association of Crime Lab Directors has begun to implement proficiency exams for anyone providing DNA forensic analysis.[48] Medical diagnosis, though, may become the most lucrative market of all. There is an enormous effort to develop tests for predisposition to cancer; a recent book on the topic predicts that the development of such tests "will have major commercial significance."[49]

Big dollars always raise questions of who will benefit; more attention needs to be given to who is likely to profit from genomics and how. Will there be differential access to genetic services? Who is going to pay for counseling? Access to medical services in the United States is already profoundly unequal;

one can imagine a situation where the well-to-do avail themselves of genetic services while the poor suffer the burden of genetic disease untreated. More to the point: it is hard to defend the Genome Project in terms of medical returns for the dollar—if that is what is intended. Dollar per dollar, if the U.S. government wants to improve American health, dollars would be better spent on providing prenatal care for the country's poor. Health care critics have long argued that public health measures are generally more cost-effective than basic research; prevention is more powerful than cure.[50] Food and drug quality enforcement, pollution prevention, and occupational health and safety regulations are more likely to maintain health for larger numbers of people than genetic research. Genomics may well become a source of improved therapies, but policy makers' priorities are not what they should be.

Social Stigma

Bioethicists—the new professional elite whose expertise is called upon in such matters—commonly suggest that genetic therapy should be used to treat or prevent illness but not to improve our genetic heritage.[51] In an older jargon, "negative eugenics" is preferable to "positive eugenics." Hormonal or genetic implants may be used to treat dwarfism, but not to increase the height of people of average stature. A related rhetoric asks that the goal of therapy be always for the individual, not for the population.

On the matter of therapy for the individual versus population therapy, one can easily imagine that the effect of genetic testing (and therapy) is likely to be populational, even if the motivation is individual. The incidence of both Tay-Sachs disease and beta-thalassemia, for example, has already dropped in response to genetic testing among the ethnic populations known to be at risk. Since testing for Tay-Sachs began in the 1970s the number of babies born with the disease has declined to only about one-tenth of previous levels.[52] One can surely expect similar results for other diseases: populational shifts will eventually result from individual decisions.

With regard to the question of negative versus positive eugenics, it is important to recall, with Thorstein Veblen, that invention is often the mother of necessity. Social pressures can create new needs. Once genetic supplements become available enabling us to alter our strength, height, sex, rate of aging, or susceptibility to disease, for example, how much pressure will there be to have our children take part? In the 1960s, orthodontics became a status symbol and crooked teeth a sign of poverty; will the same become true for other, once inalterable, traits?[53]

From a practical point of view, questions such as these will probably be phrased in the form: which among the many dozens of possible tests should a pregnant woman—or a couple seeking counseling—be encouraged to take? In the fall of 1989 the American Society for Human Genetics voted to support a moratorium on widespread testing for cystic fibrosis, fearing a rush of millions of people to have themselves tested (twelve million Americans carry a single copy of the gene; one in 2,000 babies inherits two copies and comes down with the disease).[54] Prohibitive costs may keep certain tests out of most people's reach—a single test for Huntington disease, for example, costs about $8,000, excluding counseling[55]—but even cheaper procedures may arouse official ire if the motives behind such tests are judged to be suspect. In 1971 the government of India banned abortions for the purpose of sex selection (one study showed that in Bombay, 99 percent of all abortions were of female fetuses); in 1988 the state of Maharastra introduced legislation to bar the use of amniocentesis for the purpose of selecting sex.[56]

Stigma in the most serious cases may bring with it legal obligations. U.S. courts recently recognized the "right to reproduce" as part of an individual's right to privacy; such rights extend to matters of contraception and abortion, as well as the right to refuse medical treatment or sterilization. Still, a person may be held legally responsible for failing to protect the genetic health of his or her offspring. Physicians who fail to properly counsel parents on the likelihood of giving birth to children with genetic lesions may be sued for "wrongful birth"; patients have been found to have the right to be fully informed of the likelihood of such lesions and what might be done to prevent them.[57] California legislators have required that physicians offer tests for maternal serum alpha-fetoprotein to pregnant women (to reveal neural tube defects); women can avoid the test only by signing a waiver. With the routinization of genetic testing (amniocentesis and chorionic villus sampling are the most common methods), new questions are sure to arise. Can parents be convicted of negligence for failing to obtain genetic therapy for their children? What if they fail to provide a special diet for a newborn diagnosed as having a life-threatening enzyme deficiency?

The stigma against genetic disease may result in an extension of coercive powers of public health. State and federal laws already exist requiring notification of many health conditions, including not just birth and death, but gunshot wounds, venereal disease, and numerous other contagious maladies. Texas legislators in 1989 sought to make it a crime for women testing positive for AIDS to bear a child; the bill was defeated, but similar efforts may arise in the case of genetic disorders. Laws in several states already require that newborns undergo blood enzyme tests for as many as nine genetic disorders (PKU, for example); physicians who fear lawsuits alleging a "failure to inform" may begin to require other tests. Widespread fears of genetic disease could conceivably result in an extension of medical authority over decisions to marry or have children, although most U.S. courts would probably be reluctant to uphold strong bans (marriage between first cousins, for example, is illegal in all U.S. states except Rhode Island and Alabama). Genetic tests are certainly not going to be used only in the countries of their discoverers, however, and it is difficult to predict how other societies will wield these tools.

Discrimination in Jobs and Insurance

Will employers try to screen employees to eliminate persons susceptible to specific genetic infirmities? What rights will insurers have to genetic information on their clients, and what rights will persons with "genetic lesions" have to keep that information private?[58]

Concerns about screening have been voiced at least since the XYY scare in the 1960s, when it was "discovered" that males with an extra Y chromosome tended to show criminal behavior. Calls to screen infants for the criminal gene were met by protests that this would stigmatize persons, even though there was only ambiguous evidence that the anomaly led to criminality (some suggested that the higher rate of convictions could come from lower intelligence). Concerns about screening also harken back to the case of sickle cell anemia. In the 1970s, a number of states initiated sickle cell screening in the black community after publicity surrounding the deaths of four black American Army recruits found to have sickling of their blood cells. A National Academy of Sciences committee recommended that recruits be screened and carriers be barred from duty as pilots (the four recruits had died while training at the moderately high-altitude Fort Bliss in Texas). The U.S. Air Force Academy for six years barred carriers of the sickle cell gene (regardless of whether they showed evidence of the disease); the Department of Defense initiated a policy of excluding carriers from aviation and flight crew training. Several corporations introduced screening amid worries that carriers might perform poorly or absorb medical resources. J. R. Murphy had shown as early as 1973 that 7 percent of all black members of the National Football League were carriers of the gene, with no apparent diminishment of performance, even in the high altitudes of the Denver stadium. Protests that blacks were being unfairly singled out as a "diseased race" eventually put a halt to widespread screening, although not before charges of discrimination, eugenics, and even genocide had been raised.[59]

Insurability raises another set of concerns. In the United States, most insurance companies are profit-making corporations, not public services or utilities. Profit for an insurance company means insuring the healthy at cheaper rates than the ill or those likely to become ill. Insurance companies therefore discriminate according to age, sex, occupation, place of residence, blood cholesterol, blood pressure, and so forth. Elaborate epidemiological tables are constructed to determine who is a good risk and who is a bad one—insurance rates are generally lower if you do not smoke or drink, just as your car insurance is lower is you have few traffic violations.* A positive test for AIDS, for example, is a virtual guarantee that one cannot obtain health or life insurance, except at prohibitive rates. The Office of Technology Assessment estimates that every year in the United States, 164,000 applicants for individual insurance policies are turned down for medical reasons.

Genetic data will probably soon become part of the information companies rely on to determine insurance rates. Companies may initially look only for perhaps half a dozen of the leading killers (for example, Huntington disease, cystic fibrosis, Tay-Sachs disease, sickle cell anemia, Alzheimer disease); eventually, with diagnostic costs declining, the repertoire of tests could include dozens of diseases. Critics charge this will lead to discrimination, and indeed it will. But insurance companies are already able to discriminate on the basis not just of sex and county of residence, but also of personal factors such as age, gender, and obesity; companies routinely ask whether you smoke or use alcohol, the nature of your work, whether you have diabetes, and so forth. Employees in businesses involving hazardous or seasonal work (in mines, lumberyards, munitions plants, or asbestos-related industries, for example, and in hotels, restaurants, or pest-control services) often find it difficult to obtain insurance.[60] Companies are legally barred from discriminating on the basis of race, but this is easily circumvented by tying rates to place of residence. Analysis of such risks leads a company to decide on the rates a person will be charged. From the insurer's point of view, it is precisely this discrimination that allows companies to provide insurance at affordable rates.[61]

One alternative to discrimination in jobs or insurance are laws governing the gathering or release of genetic data. The U.S. Supreme Court in March 1991 ruled that employers could not exclude women from certain jobs simply because they (or their future children) were at risk from exposure to workplace hazards (lead, in particular); similar rulings could be used to bar other kinds of genetic discrimination.[62] The Americans with Disabilities Act prohibits companies from requiring the traditional preemployment medical examination; it is not yet clear how effective this will be in barring genetic discrimination. In an ideal world, of course (the kind of world most Euro-

*Some of the first epidemiological studies—of hazards in the mining industry, for example—were done by insurance companies trying to determine which industries should be insured and which should not. Insurance companies in the early part of this century were champions of occupational health and safety reform, often linking their promises to insure a factory or a mine's workers with requirements that the industry clean up its act or else lose its policy.

peans have enjoyed for decades), the obvious alternative to profit-maximizing private insurers is some form of national health care. If society as a whole can assume the financial burden for medical care, then one need not ever see the growth of an unemployable, uninsurable "genetic underclass."[63]

Military Applications

Part of the original appeal of the Genome Project was that, unlike big science projects in physics, there appeared to be little possibility of military gain from the project. In June 1986, at a Cold Spring Harbor symposium on "The Molecular Biology of *Homo sapiens*," the British geneticist Walter Bodmer concluded his keynote address by arguing that the project was both "enormously worthwhile" and had "no defense implications."[64]

Molecular biology would be a very unique field indeed if there were no possibility of military use or abuse. In fact, the U.S. military command has long been interested in potential military applications of genetic knowledge, as part of its larger interest in questions of "biodefense."[65] Military exploitation of recombinant DNA technologies began in the 1970s, with efforts to design vaccines that could be used to protect friendly forces against an offensive enemy biowarfare attack. Military support for the new technology culminated in 1986, when $90 million was spent on Reagan's biodefense program.

Although most of the research in this area is classified, a number of authors have provided some sense of the kinds of goals and capabilities one could expect of military biotechnology. Genetic manipulation can increase the hardiness or virulence of a germ, allowing it to survive aerial spraying or to deliver more powerful and deadly toxins. Bacterial agents can be made resistant to antibiotics or invisible to diagnostics, making it difficult for an enemy to develop countermeasures. Genetic manipulation can also make it easier to study and produce such agents. Monoclonal antibodies can be used to produce large volumes of specific and very pure antibodies; recombinant DNA technology can be used to develop strains of agents that are "safe" during laboratory manipulation.[66]

Some of the suggested possibilities of military biotechnology sound like science fiction. It might be possible, for example, to engineer the genome of a virus to make "mistakes" as it replicates, making it difficult for the body to develop resistance. (Such a possibility no doubt led to—unfounded—Soviet suspicions, in the mid-1980s, that the AIDS epidemic might have begun from an accidental release of a biowarfare agent from U.S. military labs.) Human populations differ in susceptibility to certain forms of disease, and for a while there was talk about the prospects of an "ethnic bomb," designed to infect one race while leaving another intact. (Valley fever, for example, is a fungal disease studied by the Navy; some studies suggested that blacks are far more vulnerable to the disease than are whites.)[67] Such ideas may or may not prove fanciful; failure to develop such weapons, though, will most likely not be from a lack of trying. It is not hard to dream up other scenarios—experts are often reluctant to speculate about possibilities in this area, fearing their ideas will be taken seriously in military labs.

The most common rationale for biowarfare research is that such research is needed for defensive purposes. Foreign governments are presumably developing such agents: shouldn't the United States be prepared to counter that threat? Critics respond that it is not always so easy to distinguish offensive and defensive capabilities. A vaccine designed to protect friendly troops could also be used to protect one's own population during an attack on enemy forces. Biowarfare defenses may reduce an aggressor's fears of retaliation or backfire; biodefense may in this sense be destabilizing in the same way that antiballistic missile defense systems are destabilizing.

It is difficult to predict precisely how knowledge of human nucleotide sequences will be used for military purposes. Apart from the obvious problem of secrecy, it is difficult even to obtain funding for (nonmilitary) research on the ethics and social implications of such activities. The National Science Foundation's program on "Ethics and Values in Science and Technology," the natural source of funding for such studies, explicitly bars research "focussing solely or primarily on ethical issues associated with military technology and national defense strategies."[68] Whether the NIH ELSI group tackles this problem remains to be seen. If, however, as some predict, biology supersedes physics as the "science of the twenty-first century," and if the militarization of science continues unchecked (nearly 70 percent of all U.S. federal research and development funding presently goes to the Department of Defense), then one can certainly expect the science of life to assist in the science of death.

Biological Determinism

Central to eugenics ideology was the view that biology is destiny—that human talents and institutions are largely the product of our anatomical, neurological, hormonal, genetic, or racial constitution. Eugenicists exaggerated the role of genetics predisposing one toward a life of alcoholism, crime, or other human defects or talent. At the root of the movement was a set of fears: that the poor were outbreeding the rich, that feminists were having too few babies, that the mentally ill and criminal were about to swamp the superior elements of the population with their high birth rates. Eugenics policies were designed to combat those fears.

Among all the potential dangers of human genomics, to my mind the most all-encompassing is the danger of its confluence with a growing trend toward biological determinism. Biological determinism is the view that the large part of human talents and disabilities—perhaps even our tastes and institutions—is anchored in our biology. The Human Genome Project has already been criticized by groups who fear that the ultimate rationale for the project is a biological determinist one. James Watson did little to dispel this concern, defending the Project as providing us with "the ultimate tool for understanding ourselves at the molecular level. . . . We used to think our fate was in our stars. Now we know, in large measure, our fate is in our genes."[69] Critics point to the long, seamy tradition of eugenicists exaggerating the role of genes in human behavior; even without the impositions of a heavy-handed state, there are dangers of seeing biology as destiny. Genes have become a near-universal scapegoat for all that ails the human species. Even where genetic influence is well established, critics worry that aggressive promoting of genetic testing may generate fears out of proportion to actual risks. In the rush to identify genetic components to cancer or heart disease or mental illness, the substantial environmental origins of those afflictions may be slighted.

I want to emphasize this danger of exaggerating the role of genetics in the development of disease. Take the example of cancer. A number of rare cancers are known to be the result of heritable genetic defects. Highly heritable cancers include retinoblastoma (associated with a deletion in chromosome 13), certain leukemias, xeroderma pigmentosum (which predisposes one toward skin cancer), and a number of rare malignancies associated with a deletion in the recently discovered p53 gene (linked to the Li-Fraumeni syndrome). These are all germline defects—in other words, the genes causing these cancers can be passed from generation to generation.[70] A number of genes have also been found that predispose the carrier to more common kinds of cancers. In the spring of 1991 a research team headed by Bert Vogelstein at Johns Hopkins announced the discovery of a tumor suppressor gene on chromosome 5, the deletion of which seemed to be implicated in the onset of colon cancer.[71] A great deal of research has also gone into the search for predisposing genes for breast cancer.[72] E. B. Claus, N. Risch, and W. D. Thompson estimate that as many as 5 to 10 percent of all cases of breast cancer can be accounted for by inherited factors, and that the distribution of breast cancers is consistent with the existence of an autosomal (non-sex-linked) dominant gene affecting about one-third of 1 percent of all women.[73] Still other studies have tried to demonstrate differential susceptibilities to lung cancer. Smoking is clearly a cause of small-cell carcinoma of the lung, but only a fraction of those who smoke heavily do in fact develop lung cancer.[74] Predisposing genes may provide an answer to the question posed by *Science* magazine reporter Jean Marx, "Why doesn't everybody get cancer?"[75]

Genetic differences no doubt account for at least part of the differential susceptibility to cancer, though assertions about their frequency and what this implies for social policy are politically charged. Popularizers are fond of providing estimates for the frequency of susceptibility genes: a front-page *New York Times* article reporting on the March 1991 discovery of a tumor suppressor gene for colon cancer states that "at least 20 percent" of all colon cancers ("and possibly many more") begin through the action of the newly discovered gene.[76] Mark Skolnick, coauthor of a widely cited *New England Journal of Medicine* report on the genetics of colon cancer, speculates that genes predisposing to colon cancer are inherited "by as many as one-third of all Americans." Skolnick draws clear policy implications from the work: if suitable markers can be found, "we could use a simple blood test to screen the entire U.S. population. Those with the gene or genes would know they are carrying within them a potentially dangerous genetic defect. They would be warned to get regular checkups, and avoid the kinds of high-fat foods thought to trigger the cancer." Skolnick also states that "one-third of Americans stand some risk of developing colon cancer, while the other two-thirds probably aren't at risk at all."[77]

Quite apart from the logistical difficulties of screening "the entire U.S. population," a number of questions can be raised about such statements. For one thing, the figures commonly given for predisposing genes for cancer are somewhat speculative. The statistical models used to generate such figures (most recently for breast and lung cancers) are designed to measure the extent to which cancer runs in families,[78] but they also have notorious problems controlling for the fact that families often share common exposures to mutagens (through the "heritability" of occupation or household environment, for example). Furthermore, the genes discovered in the 1991 *Science* study headed by Vogelstein are genes in the DNA of tumor cells, not in germline DNA. The finding of damaged or deleted genetic loci in tumors does not mean that the same defects will be found in the remaining cells of the body and that the defects are therefore heritable. Finally, there is little evidence for the claim that two-thirds of Americans "probably aren't at risk at all" for a disease such as colon cancer. Such a claim presumes a small number of predisposing genes when the actual number might be quite large, producing a continuum of differential susceptibility rather than a simple yes or no, "susceptible or not." There is no evidence that a sizable fraction of the American population is invulnerable to cancer.

The most serious objection to predisposition studies, though, is that they can detract attention from the epidemiological fact that cancer is a disease whose incidence varies according to occupation, diet, socioeconomic status, and personal habits such as smoking. Genetics can do little to explain such patterns. Rates of lung and breast cancer, for example—two of the three deadliest cancers—have both risen dramatically in recent years; genetic propensities can have little to do with such increases. Lung cancer rates for U.S. males rose from 5 per 100,000 in 1930 to 75 per 100,000 in 1985—a fifteen-fold increase.[79] Lung cancer rates for women rose some 420 percent over the last thirty years, and breast cancer rates have also grown. The American Cancer Society's 1991 *Cancer Facts and Figures* estimates that American women now face a one in nine chance of developing breast cancer—up some 10 percent from the risk calculated only four years earlier.[80] Such dramatic rises lead one to suspect environmental changes, rather than genetic propensities, as the primary culprit. Even Claus and his colleagues concede that "the great majority of breast cancers are nongenetic."[81]

A similar argument can be made for most of the systematic differences in cancer rates found between ethnic groups. Blacks, for example, have significantly higher cancer rates than do whites. But, as a National Cancer Institute study revealed in the spring of 1991, poverty—not race—is the primary cause of that difference.[82] And surely there are many other differences for which genetics will be irrelevant. Genetics is not going to explain the fact that asbestos miners have higher mesothelioma rates than people who work in air-conditioned offices, nor will it explain the fact that people who live in homes with radon seepage are more likely to contract cancer. Even if individuals vary in susceptibility to such agents, it may be wishful thinking to imagine that physicians will be able to assure people, from their genetic profiles alone, that they are or are not at risk for common diseases such as cancer. Nancy Wexler, president of the Hereditary Disease Foundation and chair of the ethics group of the Human Genome Project, has recently suggested that "[a]s geneticists learn more about diabetes or hypertension or cancer, at some point they will cross an important line. Instead of saying, as they do now, 'Lung cancer runs in your family and you should be careful,' physicians will be able to ask their patients, 'Would you like to take a blood test to see if you are going to get lung cancer?'"[83] But if the majority of lung cancers are environmentally induced (and there is evidence that cigarette smoking alone accounts for as much as 90 percent of all lung cancers),[84] then physicians are unlikely ever to be able to predict cancers of this sort at an early age—except perhaps for the small percentage for whom a clear genetic defect can be discovered. It is misleading to suggest that physicians will ever have this power.[85]

Part of the confusion arises from the fact that cancers may be "genetic" in two very different senses. In one sense, all cancers are genetic. All cancer involves the runaway replication of cellular tissue; carcinogenesis invariably involves the switching on or off of genes that normally would not act in this fashion. Cellular replication involves genetic expression, and in this sense all cancers are genetic. Only a tiny proportion of cancers, however, are known to be heritable—that is, transmissible from generation to generation. There are thus two types of cancers, or rather two different ways cancers may originate: *somatic* cancers arise from genetic transformations in some particular bodily tissue (caused by exposure to mutagens or viruses, for example); *germline* or *heritable* cancers are passed from generation to generation through the genetic information in the germ cells (the sperm or eggs). The genetic defect is distributed differently in the two cases. In somatically induced cancers, the genetic malfunction lies only in the injured cells of the tumor. In heritable, germline cancers, the genetic defect will be found in every tissue of the body. The distinction is not always clear-cut: as already noted, some heritable genes confer an increased susceptibility to cancer, the ultimate trigger for which is some environmental mutagen. A given cancer (retinoblastoma, for example) may have both familial and sporadic (somatic) forms—the distinction being only in the timing and location of the mutation. The appropriate therapies for heritable and somatic cancers may be indistinguishable. Still, the root cause of cancer in the two extreme cases is quite different. Germline cancers may be expressed regardless of the environment to which one is exposed; somatic cancers are generally triggered by some postnatal environmental insult.

Knowledge of the genetic mechanisms involved in carcinogenesis has been growing rapidly in recent years. Since the late 1970s, several dozen different segments of the human genome have been discovered that, when mutated (sometimes by as little as one nucleotide, as in the case of human bladder cancer), produce cancer. "Oncogenes," as these segments are called, have been found associated with viruses that infect chickens, monkeys, and cats. Occurring naturally in certain animals, oncogenes may be picked up by these viruses and transmitted in the course of infection. Genes have also been discovered that, by contrast with oncogenes, allow cancer to flourish when the gene is absent. These "tumor suppressor genes" normally prevent the growth of cancer; when damaged or deleted, as seems to be required for the development of retinoblastoma, Wilms tumor (a kidney cancer), and certain forms of colon cancer, it is the absence or impairment of the gene that allows the malignancy to grow. How and why these growth-blocking genes are first activated or deactivated is not yet clear.[86]

The important question of policy interest, though, is: What causes a gene to mutate in the first place? Much has been made of the fact that carcinogenesis begins with genetic changes; Robert Weinberg of the Whitehead Institute (and one of the discoverers of oncogenes) states that "the roots of cancer lie in our genes."[87] Improved therapies may well emerge from investigating precisely what biochemical functions are curtailed by the loss of tumor suppressor genes. If, however, as most often appears to be the case, cancer begins through some kind of environmental insult—exposure to ionizing radiation, for example, or to one of the many chemical carcinogens in our air, food, or water—then the fact that oncogenes must be activated and suppressor genes turned off does not alter the fact that, from a long-run societal point of view, prevention is probably going to remain the best way to approach the problem of cancer.

One of the dangers of biological determinism, then, is that the root cause for the onset of disease is shifted from the environment (toxic exposures) to the individual (genetic defects). The scientific search shifts from a search for mutagens in the environment to biological defects in the individual. Geneticists sometimes argue that identification of persons especially at risk will allow us to determine "who will benefit maximally from treatments designed to manipulate the environmental causes of those conditions."[88] Critics, however, point out that susceptibilities may be used for less benevolent purposes. *Consumer Reports,* in its July 1990 cover story on genetic screening, warned, "The danger is that industry may try to screen out the most vulnerable rather than clean up an environment that places all workers at increased risk."[89]

The threat screening poses to individual workers is well known;[90] less well known may be the fact that a number of industry spokesmen have sought to play the genetic card to defend a particular product as safe. The Council for Tobacco Research has spent more than $150,000,000 on biomedical research since its founding in 1954; the overwhelming majority of its widely disseminated cancer studies have been devoted to genetic research.[91] Tobacco lobbyists have tried to argue that smoking causes cancer only in those persons for whom there is already a genetic predisposition.[92] The recent discovery of a gene triggering the onset of lung cancer (by converting hydrocarbons into carcinogens when exposed to cigarette smoke) prompted the discoverer to assert, "If we could identify those people in whom this gene is easily activated, then we could counsel them, not only not to smoke, but to avoid exposure to certain environmental pollutants."[93] Again, the danger in such arguments is that emphasis is placed on defects in the individual rather than defects in the industrial product or environment. The risk is what might be called an ideological one: if the (mis)conception grows that "nature" is more important than "nurture" in the onset of certain diseases, lawmakers may find themselves less willing to enact strong pollution prevention measures or consumer protection legislation.

It is not always easy, of course, to separate "nature" and "nurture" in such matters. Discoveries of genes for Alzheimer disease, manic depression, schizophrenia, Tourette syndrome, and lung cancer have all been announced with widespread media attention, only to be later found seriously flawed.[94] Even for diseases that are clearly heritable, genes may vary widely in their manner of expression. Five percent of men who have heart attacks before the age of sixty carry a gene that prevents the liver from filtering out harmful cholesterol. Not all of those men with the gene, however, suffer early heart disease—only about half do. Diet or some other factor can apparently ameliorate the negative effects of the disease. Neil Holtzman, professor of pediatrics at Johns Hopkins, points out that genetic tests may have little predictive value in cases where genetic expression is highly variable: "For the vast majority of people affected by heart disease, cancer and the like, the origin is so complex that it's a gross oversimplification to think that screening for a predisposing gene will be predictive."[95]

Conclusion

Popularizers and experts both often present the birth of genomics as heralding newfound technical and moral powers. The *Time* magazine cover story on the Human Genome Project of March 20, 1989, asserts that scientists are likely to be able eventually to predict an individual's vulnerability not just to diseases such as cystic fibrosis, but also for "more common disorders like heart disease and cancer, which at the very least have large genetic components."[96] Others imply sweeping moral transformations from the newfound knowledge of our nucleotides. An article in the *Hastings Center Report* envisions "a heightened societal attention to heritage, with DNA stored in banks becoming a new type of ancestral shrine"; the authors forecast a "renewed commitment to intergenerational relatedness." Daniel Koshland, in an editorial in *Science,* suggests that sequencing the human genome may result in "a great new technology to aid the poor, the infirm, and the underprivileged." George Bugliarello, of Brooklyn's Polytechnic University, states that human genome research may help us understand our "constitutional propensity to violence and aggression"; genome research is supposed to give us, for the first time, "a serious if distant hope of finding ways to change some of our dangerous ancestral traits."[97]

At the risk of dashing such fanciful hopes, it is important to keep in mind that there are not that many diseases that have a clear and simple genetic origin. Cystic fibrosis kills on the order of 500 Americans per year; death rates from Huntington disease are roughly comparable. These are not insignificant numbers, but they must be put into perspective. Cigarette smoking alone, for example, is estimated to cause the deaths of some 400,000 Americans per year—more, in other words, than all known genetic diseases combined.[98] Heart disease takes an even higher toll. If it is lives we want to save, where is the $3 billion antismoking initiative or the billion-dollar effort to reduce saturated fats or food additives? Where are the billions to reduce occupational exposures to toxins, or radon gas in homes? The U.S. infant mortality rate is among the worst for industrial nations; knowledge of genetics will do little to help change that. If improved health is our goal, then surely there is something wrong with the priorities of medical funding.

There is a great deal of concern about the "social control" likely to flow from the Human Genome Project. The argument I've made here is that there is an equal danger in the *illusion of control* that will flow from people assuming that everything is genetic. We are in the midst of an upsurge in biological determinism, and not just in areas such as cancer theory. There is little danger today of the kinds of abuses that 1920s and 1930s eugenicists foisted on the world. The emphasis today is on treating or preventing genetic disease, not on elevation of the health of the race. Efforts are aimed at voluntary therapy, not forcible sterilization or marital bans. Counseling is supposed to be self-consciously "nondirective," meaning that it is left to individual patients to decide what kind of therapy they (or their offspring) will or will not have. Unlike in the eugenics movement, those who have genetic diseases are often those in the vanguard pushing for therapy. The champions of genomics are among those calling for research into the social and ethical implications on the project: the NIH working group formed to monitor the "Ethical, Legal, and Social Issues Related to Mapping and Sequencing the Human Genome" has already begun to fund research into most of the concerns outlined here, as well as several others.[99] The DOE has begrudgingly agreed to support ethics research.[100] Most important, perhaps, is the fact that American society as a whole has changed. Civil rights advocates have successfully pushed for laws protecting the status of minorities; disability movements have resulted in guarantees of access for the handicapped. The increasingly powerful voices of women and minorities have made it harder to stigmatize groups in the fashion of the 1920s and 1930s. Times have changed.

The biological determinism that underlay the 1930s eugenics movement, however, has by no means disappeared. Genetics remains very much a "science of human inequality" insofar as the more we look for differences, the more likely we are to find them. In the face of unequal powers and unequal access, there is a great danger of exaggerating the extent to which human behavior is rooted in the genes. Scientists still work to prove that intelligence, alcoholism, crime, depression, homosexuality, female intuition, and a wide range of other talents or disabilities are the inflexible outcomes of human genes, hormones, neural anatomy, or evolutionary history.[101] Endocrinologists still prescribe hormonal therapies to prevent homosexuality; psychologists still try to prove that average differences in black-white IQ scores are genetic. It is not so long since sociobiologists suggested that women are unlikely ever to achieve equality with men in the spheres of business and science.[102]

Biology is a common and convenient explanation for intractable social problems. In 1979, amid growing fears of international terrorism, *Science* magazine reported research claiming that "most terrorists probably suffer from faulty vestibular functions in the middle ear." In 1989, with violence growing in the schools, the physician Melvin Konner wrote in the *New York Times* that the tendency for people to do physical harm to others was "intrinsic, fundamental, natural."[103] Genetics has been blamed for nearly every conceivable vice and folly of human life; in 1991, scientists from Penn State and the University of Colorado's Institute for Behavioral Genetics announced that television viewing habits (including amount watched and even preferences for particular kinds of shows) were rooted in the genes.[104] Scientific pronouncements on such issues are regularly picked up by the popular press: in 1978, at the height of the sociobiology controversy, *Playboy* magazine announced that scientists had found both rape and infidelity throughout the animal kingdom. The title of the article said it all: "Do Men *Need* to Cheat on Their Wives? A New Science Says YES: Darwin and the Double Standard."[105]

Critics worry that science in such cases is being used as a proxy for deeply held social values—that women cannot compete, that blacks are inferior, that war or crime or homosexuality or poverty is a disease that must be combated by medical means. Critics point out that there is little evidence that terrorism, or sexual preference, or personality traits such as "shyness" or "bullying" are genetically anchored, and that it is easy to mistake the intransigence of human cultural qualities (aggression or rape, for example) for biological anchoring.[106]

If there is a disconcerting continuity between genomics and eugenics, it is the fact that both have taken root in a climate where many people believe that the large part of human talents and disabilities are heritable through the genes. The study of human biological differences is not an inherently malevolent endeavor,[107] but in late twentieth-century America, as in midcentury Germany, it is dangerous to assume that biology is destiny. Genetic disease is a reality, but the frequency of such disease should not be exaggerated. From the point of view of lives saved per dollar, monies would probably be better spent preventing exposures to mutagens, rather than producing ever more precise analyses of their origins and effects. Sequencing the human genome may be a technological marvel, but it will not give us the key to life. Pronouncements that "our fate is in our genes" may be good advertising for congressional support, but they may well exaggerate the benefit that will flow from knowledge of our nucleotides. The genome is not "the very essence" of what it means to be human, any more than sheet music is the essence of a concert performance.

But criticism must be concrete, not abstract. Critics sometimes warn about the "slippery slope" of technological development—that if you allow x, then what is to prevent you from doing y? The whole purpose of law, as of ethics, is to draw lines through the continuum of social space—lines that limit what can or should be done, independent of nature's continuum. Criticism of technologies must be rooted in understanding how specific harms and benefits are distributed over particular social groups; it is as easy (and as mindless) to slide down the slippery slope of criticism, condemning any and all manipulations as "playing God," as it is to plunge headstrong across new technological frontiers.

Optimists might like to imagine that old-style abuses are a thing of the past, but a bit of perspective reveals otherwise. Singapore has launched a eugenics program that rivals those of the 1930s; China in 1989 began a systematic effort to sterilize its mentally retarded.[108] In the United States many safety nets are in place, but new forms of technological acrobatics create ever-new pits into which people are likely to fall. We should keep in mind that the potential for abuse of any technology is largely dependent on the social context within which that technology is used. This is the kernel of truth behind the claims that the Human Genome Project is not likely to raise any new ethical questions. The questions most commonly raised concerning discrimination, unequal access to resources, knowledge in the face of impotence, and rationalizations of social inequality are all old ones in medical ethics. Abuses stem from powers unequally distributed. The danger, to my mind, is therefore not that people will try to improve the genetic health of their offspring, or even that the health of groups as well as individuals will be the target of health planners. The danger is that in a society where power is still unequally distributed between haves and have-nots, the application of the new genetic technologies—as of any other—is as likely to reinforce as to ameliorate patterns of indignity and injustice.

NOTES

1. L. Jaroff, "The Gene Hunt," *Time,* March 20, 1989, p. 6.
2. U.S. Department of Health and Human Services and U.S. Department of Energy, *Understanding Our Genetic Inheritance. The U.S. Human Genome Project: The First Five Years FY 1991–1995,* National Technical Information Service, Springfield, VA, 1990. Early plans for the project are described in the National Research Council's report *Mapping and Sequencing the Human Genome,* National Academy Press, Washington, D.C., 1988; compare also the

Office of Technology Assessment's *Mapping Our Genes: The Genome Projects: How Big, How Fast?*, Government Printing Office, Washington, D.C., 1988, and J. D. Watson, "The Human Genome Project," *Science* 248:44–51, 1990.

3. Popular accounts of the project include: J. E. Bishop, M. Waldholz, *Genome: The Story of the Most Astonishing Scientific Adventure of Our Time,* Simon & Schuster, New York, 1990; J. Davis, *Mapping the Code,* Wiley, New York, 1990; L. Wingerson, *Mapping Our Genes,* Dutton, New York, 1990; R. Shapiro, *The Human Blueprint: The Race to Unlock the Secrets of Our Genetic Script,* St. Martin's Press, New York, 1991; compare also the stinging review of these works by D. B. Paul in *Science* 252:142–3, 1991. More scholarly accounts are: L. Hood, D. Kevles, eds., *The Code of Codes: Scientific and Social Issues in the Human Genome Project,* Harvard University Press, Cambridge, MA, 1992; T. F. Lee, *The Human Genome Project, The Quest for the Code of Life,* Plenum Press, New York, 1991; R. Cook-Deegan, *Gene Quest: Science, Politics, and the Human Genome Project,* Norton, New York, 1993 (in press).

4. R. L. Sinsheimer, letter to *Science* 249:1359, 1990.

5. G. Spears, "Did Lincoln Have a Disease?" *Philadelphia Inquirer,* May 3, 1991.

6. T. Appenzeller, "Democratizing the DNA Sequence," *Science* 247:1030–2, 1990. For a science fiction presentation of the possibilities of molecular paleontology, see M. Crichton, *Jurassic Park,* Knopf, New York, 1990.

7. In their eagerness to obtain "the" genome map, molecular geneticists have tended to ignore human genetic variability. One of the more exciting applications of sequence data, however, will hopefully be the enlarging of our ideas about human relatedness and variability. See L. Roberts, "A Genetic Survey of Vanishing Peoples," *Science* 252:1614–6, 1991.

8. On October 12, 1990, *Science* magazine published a glossy outline map of the human genome listing the diseases known as of July 1990. More than 5,000 genes have been entered into the registry Victor McKusick maintains at Johns Hopkins; see his *Mendelian Inheritance in Man,* 9th ed., Johns Hopkins University Press, Baltimore, 1991.

9. R. S. Boyd, "Science Finds Faulty Genes, but Not Cures," *Centre Daily Times,* December 9, 1990.

10. The first experiment in human gene therapy (somatic, not germline) was begun on May 22, 1989, when physicians at the National Institutes of Health (NIH) injected a terminally ill cancer patient with white blood cells into which a foreign gene had been inserted. The procedure was not, strictly speaking, "therapy" because the implanted gene was simply a marker used to trace the cells during conventional cancer treatment. See B. J. Culliton, "Gene Test Begins," *Science* 244:913, 1989. On January 29, 1991, Steven Rosenberg at the National Cancer Institute injected melanoma patients with cells into which a gene for tumor necrosis factor had been inserted. Other experiments are now underway to introduce genes that may combat AIDS, cancer, and other diseases. On the early background to gene therapy, see the Office of Technology Assessment's report, *Human Gene Therapy, Background Paper,* Government Printing Office, Washington, D.C., 1984. Somatic gene therapy involves the insertion of genetic material into cells of the body for the duration of a person's lifetime; germline therapy would involve the introduction of heritable genetic material into the sperm, egg, or nucleus of the fertilized egg. Genes introduced into germ cells would be inher-

ited in subsequent generations of the organism. Physicians are not seriously entertaining the prospect of germline therapy; most bioethicists consider it unethical.

11. "Statement of the First Workshop of the Joint Working Group on the Ethical, Legal and Social Issues Related to Mapping and Sequencing the Human Genome," Williamsburg, February 5–6, 1990, in the HHS and DOE booklet, *Understanding Our Genetic Inheritance,* pp. 65–73.

12. See, for example, D. Nelkin, and L. Tancredi, *Dangerous Diagnostics: The Social Power of Biological Information,* Basic Books, New York, 1989.

13. F. Galton, *Inquiries into the Human Faculty,* Macmillan, London, 1883, p. 44.

14. In 1895 the German physician Alfred Ploetz published *Die Tüchtigkeit unsrer Rasse und der Schutz der Schwachen,* coining the term "racial hygiene" *(Rassenhygiene)* to designate the study of "the optimal conditions for the maintenance and development of the race." The terms "eugenics" and "racial hygiene" are roughly synonymous—though a number of theorists in the Nazi movement tried to drive a wedge between home-grown *racial hygiene* and the more internationalist *eugenics.*

15. On Nazi genetics, see R. N. Proctor, *Racial Hygiene: Medicine under the Nazis,* Harvard University Press, Cambridge, MA, 1988; also, B. Müller-Hill, *Murderous Science: Elimination by Scientific Selection of Jews, Gypsies, and Others, Germany 1933–1945,* Oxford University Press, Oxford, 1988.

16. C. W. Saleeby, "Preventive Eugenics—The Protection of Parenthood from the Racial Poisons," in *Eugenics in Race and State,* vol. II, Williams & Wilkins, Baltimore, 1923, p. 309; also, Proctor, *Racial Hygiene,* pp. 238–9.

17. O. von Verschuer, *Erbpathologie: Ein Lehrbuch für Ärzte und Medizinstudierende,* 2d ed., T. Steinkopff, Dresden and Leipzig, 1937.

18. See, for example, *Der Erbarzt* ("The Genetic Doctor"), published as a supplement to the *Deutsches Ärzteblatt* beginning in 1934; also *Fortschritte der Erbpathologie* ("Progress in Genetic Pathology"), whose first issue appeared in 1937. See also Proctor, *Racial Hygiene,* pp. 104–8, 194–8, 212–7.

19. See Proctor, *Racial Hygiene,* pp. 97–101; also, R. N. Proctor, "Eugenics among the Social Sciences: Germany and the U.S.," in *The Estate of Social Knowledge,* edited by J. Brown and D. K. van Keuren, Johns Hopkins University Press, Baltimore, 1991.

20. See Proctor, *Racial Hygiene,* p. 348, note 99.

21. See R. N. Proctor, "From *Anthropologie* to *Rassenkunde:* Concepts of Race in German Physical Anthropology," in *Bones, Bodies, Behavior: Essays on Biological Anthropology,* edited by G. Stocking, University of Wisconsin Press, Madison, 1988.

22. See Proctor, *Racial Hygiene,* pp. 298–308; also Proctor, "From *Anthropologie* to *Rassenkunde,*" pp. 138–79.

23. B. Müller-Hill, "Genetics after Auschwitz," *Holocaust and Genocide Studies* 2:3–20, 1987.

24. W. Kaempffert, "Bombs and People: New Power Called Compulsion to Practice of Eugenics," *New York Times,* September 30, 1945. Kaempffert curiously felt obliged to point out that, despite his concern, Dr. Burch had "no thought of killing off the socially unfit."

25. "Dr. Muller and the Million Human Time Bombs," *Science Illustrated,* May 4,

1949, p. 60; compare also J. Beatty, "Genetics in the Atomic Age: The Atomic Bomb Casualty Commission, 1947–1956," in *The Expansion of American Biology*, edited by K. Benson et al., Rutgers University Press, New Brunswick, NJ, 1991.

26. M. S. Lindee, "Mutation, Radiation, and Species Survival: The Genetic Studies of the Atomic Bomb Casualty Commission in Hiroshima and Nagasaki," Ph.D. thesis, Cornell University, 1990.

27. G. Allen, "Eugenics," *International Encyclopedia of the Social Sciences,* Macmillan, New York, 1968, pp. 193–5.

28. B. J. Barnhart, "The Department of Energy (DOE) Human Genome Initiative," *Genomics* 5:657, 1989.

29. R. Mullan Cook-Deegan, "The Alta Summit, December 1984," *Genomics* 5:661–663, 1989. The Alta meeting itself followed an earlier meeting in Hiroshima (in March 1984), where DNA analytical tools were ranked along with establishing cell lines from bomb survivors and their children as the two most promising directions for human mutation research.

30. OTA, 1988, p. 190. The FBI plans to establish a national computerized data bank of DNA patterns for convicted felons; see Eric S. Lander, "Research on DNA Typing Catching Up with Courtroom Application," *American Journal of Human Genetics* 48:819–23, 1991.

31. Cook-Deegan, 1993.

32. See N. Angier, "Vast, 15-Year Effort to Decipher Genes Stirs Opposition," *New York Times,* June 5, 1990; also Cook-Deegan, 1993. In 1987 Charles DeLisi convinced the DOE to begin its own genome program. As former head of the DOE's Office of Health and Environmental Research, DeLisi had become convinced of the need for a sequencing effort in 1985 after reading an Office of Technology Assessment report on *Technologies for Detecting Heritable Mutations in Human Beings;* see OTA, 1988, pp. 99–102, for DeLisi's recommendations.

33. Domenici and Botstein are both cited in Cook-Deegan, 1993.

34. U.S. Department of Energy, Office of Health and Environmental Research, *Human Genome: 1989–90 Program Report,* National Technical Information Service, Springfield, VA, 1990, p. 9.

35. J. Walsh, J. Marks, "Sequencing the Human Genome," *Nature* 322:590, 1986; B. D. Davis and colleagues, "The Human Genome and Other Initiatives," *Science* 249:342–3, 1990.

36. See Jeremy Rifkin's much and rightly maligned *Algeny: A New Word—A New World,* Viking Press, New York, 1983; also his earlier book with Ted Howard, *Who Should Play God?* Delacorte Press, New York, 1977. Compare Rochelle Green's cover story in *Health* magazine: "Tinkering with the Secrets of Life," January 1990, pp. 46ff.

37. P. Elmer-Dewitt, "The Perils of Treading on Heredity," *Time,* March 20, 1989, p. 70.

38. Antiabortionists commonly invoke the Nazis as part of their attack on today's supposed "holocaust" of abortions. Typical is W. Brennan's *The Abortion Holocaust: Today's Final Solution,* Landmark, St. Louis, 1983.

39. "American Society of Human Genetics Statement on Clinical Genetics and Freedom of Choice," *American Journal of Human Genetics* 48:1011, 1991.

40. N. S. Wexler, "The Oracle of DNA," in *Molecular Genetics in Diseases of Brain,*

Nerve, and Muscle, edited by L. P. Rowland et al., Oxford University Press, Oxford, 1989, p. 438.

41. Ibid., p. 430.
42. See Ray White's essay in this volume, Chapter 10.
43. S. Tanchinski, "Boom and Bust in the Bio Business," *New Scientist,* January 22, 1987, p. 47. A related problem concerns the exploitation of Third World genetic reserves. First World biologists are busily combing Third World forests in search of genetic materials for pharmaceuticals and agricultural purposes, but who owns these genetic resources? Tropical forests and savannahs have been the source of most of the world's food varieties—do First World biologists have the right to exploit these reserves without compensating the nations from which they are taken?
44. E. L. Andrews, "Patents: Applications for Animals up Sharply," *New York Times,* April 22, 1989. Patent applications have been filed for more than a hundred other genetically engineered organisms, including: a mouse carrying a gene to create human insulin (Massachusetts General Hospital); a mouse carrying a transplanted gene for AIDS (National Institute for Allergies and Infectious Diseases); a hog that produces pork with low cholesterol (Embryogen); sheep engineered to secrete blood coagulation proteins (Pharmaceutical Protein Ltd., England); cows that produce milk with higher concentrations of casein (University of Wisconsin); and many others.
45. R. S. Eisenberg, "Patenting the Human Genome," *Emory Law Journal* 39:721–45, 1990, and Chapter 14.
46. E. L. Andrews, "U.S. Seeks Patent on Genetic Codes, Setting Off Furor," *New York Times,* October 21, 1991, p. 1.
47. L. Roberts, "Ethical Questions Haunt New Genetic Technologies," *Science* 243:1134–6, 1989.
48. Lander, 1991, p. 822.
49. See M. A. Rothstein's chapter in *Detection of Cancer Predisposition: Laboratory Approaches,* edited by L. Spatz et al., March of Dimes, White Plains, NY, 1990. Expectations are that the knowledge and techniques gained through the Human Genome Initiative will prove valuable for the development of domestic animals and crop plants, although separate initiatives have also been launched in agricultural genomics. In 1989 the U.S. Department of Agriculture announced a Plant Crop Initiative to map and sequence the genomes of several of the world's major food crops (corn, wheat, and barley, among others); there are also hopes for an Animal Genome Project. See L. Roberts, "An Animal Genome Project?" *Science* 248:550–2, 1990. The Japanese have announced plans to spend $200 million mapping and sequencing the rice genome.
50. See T. McKeown, *The Role of Medicine: Dream, Mirage, or Nemesis?* Princeton University Press, Princeton, NJ, 1979; D. P. Burkitt and H. C. Trowell, eds., *Refined Carbohydrate Foods and Disease: Some Implications of Dietary Fibre,* Academic Press, London, 1975; S. Epstein, *The Politics of Cancer,* Sierra Club Books, San Francisco, 1978.
51. This was the conclusion of the President's Commission for the Study of Ethical Problems in Medicine and Biomedical and Behavioral Research; see its *Screening and Counseling for Genetic Conditions,* Government Printing Office, Washington, D.C., 1983.

52. P. R. Reilly, "Advantages of Genetic Testing Outweigh Arguments against Widespread Screening," *The Scientist,* January 21, 1991, p. 9.

53. See B. Werth, "How Short Is Too Short? Marketing Human Growth Hormone," *New York Times Magazine,* June 16, 1991, pp. 14ff.

54. L. Roberts, "To Test or Not to Test?" *Science* 247:17–9, 1990.

55. *American Journal of Human Genetics* 48:172, 1991.

56. T. Duster, *Back Door to Eugenics,* Routledge, New York, 1990, p. 34.

57. S. Elias, George J. Annas, *Reproductive Genetics and the Law,* Year Book Medical Publishers, Chicago, 1987, pp. 109–10. The "right to reproduce" remains a sensitive issue: in 1988 thousands of copies of Stephen Trombley's book *The Right to Reproduce: A History of Coercive Sterilization* were destroyed after Planned Parenthood threatened to sue the London publisher, Weidenfeld and Nicolson, for libel.

58. N. Angier, "Vast, 15-Year Effort to Decipher Genes Stirs Opposition," *New York Times,* June 5, 1990; Nelkin and Tancredi, 1989; R. Macklin, "Mapping the Human Genome: Problems of Privacy and Free Choice," in *Genetics and the Law III,* edited by A. Milunsky and G. J. Annas, Plenum Press, New York, 1985; Committee for Responsible Genetics, "Position Paper on the Human Genome Initiative," January 10, 1990.

59. Duster, 1990, pp. 24–6; Kevles, *In the Name of Eugenics,* University of California Press, San Francisco, pp. 276–9.

60. "Health Insurers are Excluding Some Groups," *New York Times,* February 5, 1990.

61. M. L. Zoler, "Genetic Tests," *Medical World News,* January 1991, pp. 32–7; R. Pokorski, "The Potential Impact of Genetic Testing," unpublished manuscript, November 30, 1989.

62. L. Greenhouse, "Court Backs Right of Women to Jobs with Health Risks," *New York Times,* March 21, 1991.

63. This has been the conclusion of a number of commentators on the Human Genome Project; see P. Thomas, "DNA D_x Invites National Insurance," *Medical World News,* September 1990, p. 29; also Nelkin and Tancredi, 1989, p. 176.

64. W. F. Bodmer, "Human Genetics: The Molecular Challenge," *Cold Spring Harbor Symposia on Quantitative Biology: The Molecular Biology of* Homo sapiens, 51:1–13, 1986.

65. L. A. Cole, *Clouds of Secrecy: The Army's Germ Warfare Tests over Populated Areas,* Rowman and Littlefield, Totowa, NJ, 1988; S. Wright, *Preventing a Biological Arms Race,* MIT Press, Cambridge, MA, 1990.

66. C. Piller, K. R. Yamamoto, *Gene Wars: Military Control over the New Genetic Technologies,* William Morrow, New York, 1988, pp. 23–5. The potential of new genetic techniques for terrorist use is substantial, given that the technologies involved are fairly low tech (by contrast with those required for nuclear capabilities).

67. Ibid., p. 24; compare also M. Lappé, *Broken Code: The Exploitation of DNA,* Sierra Club Books, San Francisco, 1984, pp. 208–9.

68. See the preface to *Ethical Issues Associated with Scientific and Technological Research for the Military,* edited by Carl Mitcham and Philip Siekevitz, New York Academy of Sciences, New York, 1989, p. xi.

69. Cited in L. Jaroff, "The Gene Hunt," *Time,* March 20, 1989:62–7.

70. R. Weiss, "Genetic Propensity to Common Cancers Found," *Science News* 138:342, 1990.
71. K. W. Kinzler et al. (including Bert Vogelstein), "Identification of a Gene Located at Chromosome 5q21 that Is Mutated in Colorectal Cancers," *Science,* 251:1366–70, 1991. B. Vogelstein et al., "Genetic Alterations during Colorectal-Tumor Development," *New England Journal of Medicine* 319:525–32, 1988; L. A. Cannon-Albright et al., "Common Inheritance of Susceptibility to Colonic Adenomatous Polyps and Associated Colorectal Cancers" in the same issue.
72. W. R. Williams, D. E. Anderson, "Genetic Epidemiology of Breast Cancer," *Genetic Epidemiology* 1:1–7, 1984; K. Wright, "Breast Cancer: Two Steps Closer to Understanding," *Science* 250:1659, 1990.
73. E. B. Claus, N. Risch, W. D. Thompson, "Genetic Analysis of Breast Cancer in the Cancer and Steroid Study," *American Journal of Human Genetics* 48:232–42, 1991. The authors estimate that 36 percent of all women with breast cancer aged twenty to twenty-nine are carriers of a gene predisposing them to the disease. Older women with the disease carry the gene with much lower frequency. The cumulative lifetime risk of developing the disease for carriers is estimated at 92 percent; noncarriers are estimated at 10 percent risk (the same as for the population as a whole).
74. N. Angier, "Cigarettes Trigger Lung Cancer Gene, Researchers Find," *New York Times,* August 21, 1990. Angier reports that only about 7 percent of those who smoke heavily develop lung cancer, but C. M. Pike and Richard Peto estimate that among men who smoke 25 or more cigarettes a day, 30 percent will develop lung cancer by age 75 in the absence of other causes of death. See *Lancet* 665–68, 1965.
75. J. Marx, "A New Tumor Suppressor Gene?" *Science* 252:1067, 1991.
76. N. Angier, "Crucial Gene is Discovered in Detecting Colon Cancer," *New York Times,* March 15, 1991.
77. Cited in Bishop and Waldholz, 1990, pp. 156, 163. Similar generalizations accompanied the November 30, 1990, publication in *Science* of evidence for predisposing genes for a rare form of breast cancer. Commenting on his discovery, David Malkin of Boston's Massachusetts General Hospital Cancer Center asserted, "We'll be able to say, 'Yes, you carry this mutation . . . and you are at risk,' or 'No, you don't.'" See R. Weiss, "Genetic Propensity to Common Cancers Found," *Science News* 138:342, 1990.
78. Claus, Risch, and Thompson's "Genetic Analysis of Breast Cancer" is a recent example; compare also T. A. Sellers et al., "Evidence for Mendelian Inheritance in the Pathogenesis of Lung Cancer," *Journal of the National Cancer Institute* 82:1272–9, 1990.
79. E. Marshall, "Experts Clash Over Cancer Data," *Science* 250:902, 1990.
80. American Cancer Society, *Cancer Facts and Figures—1991,* American Cancer Society, Atlanta, 1991.
81. Claus et al., 1991, p. 241. In one sense, of course, breast cancer is overwhelmingly a genetic disease. More than 95 percent of all breast cancers occur in individuals with the XX rather than the XY karyotype—the disease strikes women far more often than men. Skin cancer is also genetic, insofar as populations with darker skin tend to suffer less from the disease. In both cases, however, genetic factors are incidental to the ultimate cause of the disease: males rarely get breast cancer

because they don't have breasts; blacks suffer less from skin cancer because they are better protected from ultraviolet radiation. Predisposing genes for breast or skin cancer may well exist, but these are unlikely to have anything to do with either sex chromosomes or the genes controlling skin coloration.

82. "Poverty Blamed for Blacks' High Cancer Rate," *New York Times,* April 17, 1991. Adjusted for socioeconomic status, whites showed slightly higher rates than blacks for breast, rectal, and lung cancer; blacks showed greater risk for stomach, cervical, and prostate cancers.

83. Wexler is cited in R. M. Henig, "High-Tech Fortune Telling," *New York Times Magazine,* December 24, 1989, p. 20.

84. R. Doll, R. Peto, *The Causes of Cancer,* Oxford University Press, Oxford, 1981. The authors estimate that in the United States smoking accounts for about one third of *all* cancer deaths.

85. See E. S. Lander, "The New Human Genetics: Mapping Inherited Diseases," *Princeton Alumni Weekly,* March 25, 1987, pp. 10–16. Compare Natalie Angier's claim that "the study of oncogenes may not be the best hope of banishing cancer; it may be the only hope." More to the point may be the quip of Richard Rifkind, which Angier herself cites: "You want a cure for cancer? Tell the bastards to quit smoking." See her *Natural Obsessions,* Warner Books, New York, 1988, pp. 17, 141.

86. E. Fearon et al., "Clonal Analysis of Human Colorectal Tumors, *Science* 238:193–7, 1987; R. A. Weinberg, "Oncogenes and Tumor Suppressor Genes," in *Unnatural Causes: The Three Leading Killer Diseases in America,* edited by R. C. Maulitz, Rutgers University Press, New Brunswick, NJ, 1989.

87. R. A. Weinberg, "Finding the Oncogene," *Scientific American* 258:44–51, 1988.

88. The words are A. Motulsky's in the *American Journal of Human Genetics* 48:174, 1991.

89. "The Telltale Gene," *Consumer Reports,* July 1990, p. 485.

90. T. H. Murray, "Warning: Screening Workers for Genetic Risk," *Hastings Center Report,* February 1983, pp. 5–8.

91. *Report of the Council for Tobacco Research—U.S.A., Inc.,* Council for Tobacco Research, New York, 1988.

92. M. Lappé, *Genetic Politics: The Limits of Biological Control,* Simon & Schuster, New York, 1979, p. 120.

93. N. Angier, "Cigarettes Trigger Lung Cancer Gene, Researchers Find," *New York Times,* August 21, 1990.

94. In the early 1970s, for example, a number of researchers postulated that individuals might differ genetically in their ability to metabolize certain lung carcinogens. Subsequent studies, however, showed that such differences were not heritable. For a review and criticism, see B. Paigen et al., "Questionable Relation of Aryl Hydrocarbon Hydroxylase to Lung-Cancer Risk," *New England Journal of Medicine* 297:346–50, 1977.

95. Green, 1990, p. 84. See also N. Holtzman's extended discussion of these issues in his *Proceed With Caution: Predicting Genetic Risks in the Recombinant DNA Era,* Johns Hopkins University Press, Baltimore, 1990.

96. L. Jaroff, "The Gene Hunt," *Time,* March 20, 1989, p. 62.

97. K. Nolan, S. Swenson, "New Tools, New Dilemmas: Genetic Frontiers," *Hastings Center Report,* October/November 1988, p. 42; D. E. Koshland,

"Sequences and Consequences of the Human Genome," *Science* 246:189, 1989; G. Bugliarello, "The Genetic and Psychological Basis of Warfare as a Challenge to Scientific Research," in Mitcham and Siekevitz, *Ethical Issues,* p. xvi. Compare also Marc Lappé's curious speculation that genomics may allow the establishment of files of persons "genetically at risk for acquiring AIDS following infection with the human immunodeficiency virus" ("The Limits of Genetic Inquiry," *Hastings Center Report,* August 1987, p. 6).

98. A recent article in *Circulation,* the official journal of the American Heart Association, estimates that more than 50,000 Americans may be killed each year by so-called "secondhand" or "environmental" smoke. See S. A. Glantz and W. W. Parmley, "Passive Smoking and Heart Disease," *Circulation* 83:1–13, 1991.

99. The Working Group's report has been published as an appendix in the HHS and DOE's *Understanding Our Genetic Inheritance,* pp. 65–73.

100. Benjamin Barnhart was reluctant to fund ethics as part of the DOE's Human Genome Project, despite the fact that Charles DeLisi had originally expressed an interest in this area. At a December 1989 meeting of the joint DOE-NIH advisor committee for the project, James Watson cautioned Barnhart that if the DOE did not fund ethical analysis, "Congress will chop off your head" (cited in Cook-Deegan, 1993).

101. R. C. Lewontin, S. Rose, L. J. Kamin, *Not in Our Genes: Biology, Ideology, and Human Nature,* Pantheon, New York, 1983.

102. G. Dörner et al., "A Neuroendocrine Predisposition for Homosexuality in Men," *Archives of Sexual Behavior* 4:1–8, 1975; T. Bouchard et al., "Sources of Human Psychological Differences: The Minnesota Study of Twins Reared Apart," *Science* 250:223–8, 1990; E. O. Wilson, "Human Decency Is Animal," *New York Times Magazine,* October 12, 1975, p. 50. For a recent and unabashed effort to prove that the average economic underperformance of blacks (compared to whites) is traceable to differences in black-white genetic endowments, see the essays in the December 1986 *Journal of Vocational Behavior.*

103. C. Holden, "Study of Terrorism Emerging as an International Endeavor," *Science* 203:33–35, 1979; "Math Genius May Have Hormonal Basis," *Science* 222:1312, 1983; M. Konner, "The Aggressors," *New York Times Magazine,* August 14, 1988, pp. 33–4.

104. M. Gooderham, "TV Viewing Habits Seen as Hereditary: 'Couch Potatoes' Born, Study Suggests," *Toronto Globe and Mail,* June 17, 1991.

105. S. Morris, "Do Men *Need* to Cheat on Their Wives? A New Science Says YES: Darwin and the Double Standard," *Playboy,* May 1978, pp. 109ff.

106. Lewontin et al., 1983; A. Fausto-Sterling, *Myths of Gender: Biological Theories about Women and Men,* Basic Books, New York, 1985.

107. The presumption of human biological equality can be as misleading as the presumption of inequality. Take, for example, the case of cystic fibrosis (CF). Discovery of "the" CF gene was announced amid great fanfare in 1989, but it soon became apparent that while a majority of cases in the U.S. population could be traced to a simple three-base-pair deletion, the disease could also be caused by some sixty other mutations. Early hopes for a simple, comprehensive test were further set back by the discovery that human populations differ substantially according to how common the three-base-pair deletion is relative to other mutations causing the disease. Among Eastern European Jews, for example, the sim-

ple three-base-pair deletion appears to account for only about 3 percent of all CF cases. In Denmark, by contrast, the simple deletion accounts for about 88 percent of all CF cases. The usefulness of the test is therefore likely to vary greatly according to the genetic background of the population in question.

108. N. D. Kristof, "Chinese Region Uses New Law to Sterilize Mentally Retarded," *New York Times,* November 21, 1989; also his "Parts of China Forcibly Sterilizing the Retarded Who Wish to Marry," *New York Times*, August 8, 1991. For the case of India see Kaval Gulhati, "Compulsory Sterilization: The Change in India's Population Policy," *Science* 195:1300–5, 1977; for Latin America see Bonnie Mass, *Population Target, The Political Economy of Population Control in Latin America*, Latin American Working Group, Toronto, 1976; and for Singapore see C. K. Chan, "Eugenics on the Rise—Singapore," in *Ethics, Reproduction, and Genetic Control,* ed. R. Chadwick, Routledge, New York, 1987. The best recent U.S. review is Philip R. Reilly, *The Surgical Solution: A History of Involuntary Sterilization in the United States,* Johns Hopkins University Press, Baltimore, 1991.

In Britain, arguments have recently been put forward that DNA diagnostic laboratories should be judged in terms of how effectively they reduce the financial burden of caring for the handicapped. See, for example, J. C. Chapple et al., "The New Genetics: Will It Pay Its Way?" *Lancet* 1:1189–92, 1987. Angus Clarke of the University of Wales's Institute of Medical Genetics has pointed to a danger in the more general use of audits to evaluate diagnostic departments, especially insofar as "efficiency" is measured in terms of aborted fetuses per unit of diagnosed defects. The danger, as he sees it, is that cost-benefit analyses of this sort will re-introduce eugenic criteria under the guise of "social responsibility in reproduction": "It is not at all far-fetched to imagine finance for the care of the handicapped being reduced on the grounds that there will be fewer handicapped children if the genetics services operated more 'efficiently.' To encourage such services to meet targets in terms of terminations, our funding might depend upon 'units of handicap prevented,' which would pressurise parents into screening programmes and then into unwanted terminations with the active collusion of clinical geneticists anxious about their budgets. . . . There certainly is a role for public health genetics, but not for a system of eugenics by default, through the impersonal, amoral operation of a penny-pinching bureaucarcy." See his "Genetics, Ethics, and Audit," *Lancet* 335:1145–47, 1990; also his further remarks on "eugenics by accountancy" in *Lancet* 336:120, 1990.

5

The Past as Prologue: Race, Class, and Gene Discrimination

Patricia A. King

The goal of the unprecedented effort to map and sequence the human genome, spearheaded by the National Institutes of Health and the Department of Energy, is to provide an important and enduring tool for medicine and biology.[1] Thus, one of the critical justifications for the project is the promise of important diagnostic and therapeutic benefits for clinical medicine. However, genetic information generated by the project may also have adverse consequences for individuals and society. In recognition of the importance of identifying the ethical, legal, and social implications of gene mapping and developing policy options to address them, an advisory group (ELSI) has been established, and up to 3 percent of the Human Genome Project's research funds have been designated to explore these important matters. The goal is to maximize the benefits and minimize the harms flowing from information the Project generates. Although the obligation of society to take account of long-term risks and benefits that may result from the search for new knowledge has been recognized in the past,[2] this is the first time in our history that such an effort has proceeded simultaneously with the initiation of a major scientific project.

The question of justice is central to any endeavor that tries to understand the social impact of vastly increased genetic information. Who will benefit and who will be burdened as a consequence of new knowledge? This is primarily a problem of fairness and "just distribution in society structured by various moral, legal, and cultural rules and principles that form the terms of cooperation for that society."[3] In a society like ours, composed of persons from many racial and ethnic groups and economic classes, the critical question is whether the knowledge gained from the Human Genome Project will be used in a way that will benefit all.[4]

From the perspective of racial and ethnic minorities and the poor, there is reason to be skeptical about whether we are likely to be beneficiaries or victims of these gene mapping efforts. Even if we could be certain that the Human Genome Project would yield medically useful results, the central role played by race and class in our society causes some members of these vulnerable groups to withhold their applause. History teaches that one must be wary of the nation's institutions and social strategies when it comes to situations where genetic information is linked with racial, ethnic, and class differences. The nation's recent history highlights at least two distinct reasons why information generated by the Human Genome Project might not be a blessing for racial and ethnic minorities and the poor.

First, in the course of mapping and sequencing the human genome, additional correlations between genetic characteristics and racial and ethnic status are likely to emerge. This outcome is not inherently burdensome. Some commentators are optimistic that future uses of genetic information will not result in biological explanations of group differences.[5] However, genetic information historically has been used to reinforce negative stereotypes about racial and ethnic groups and poor people, rather than to deemphasize differences among groups in the United States. The eugenics movement in the United States in the early decades of the century is an excellent example of this phenomenon. Moreover, genetic information coupled with race has been used as a justification for discrimination against the members of vulnerable groups. For example, this nation's efforts to screen African-Americans for carriers of sickle cell disease has resulted in both racial stigmatization and employment discrimination.[6] The continuing debate about race and the genetic basis of intelligence is another example of the harm and stigmatization that can be the consequence of linking genetic information and race.[7]

Second, past and existing inequities in health status, health care coverage, and the delivery of health care services support fears of racial and ethnic minorities and the poor that the potential benefits of gene mapping will not be fairly distributed.[8] Members of these historically disadvantaged groups have reason to fear that they will not have access to, much less receive, the quality of genetic clinical services available to the rest of the population.

These current inequities raise questions about what strategies for health care delivery and medical research are likely to provide the maximum benefit to those who are the most disadvantaged. Even if racial and ethnic minorities and the poor were likely to benefit from gene mapping, would these groups benefit more by reallocating funds now supporting the gene mapping effort to programs that attack social conditions contributing to diseases currently posing the greatest threats to health in these segments of the population?[9] Although this cluster of issues is significant and goes to the core of the goal of providing benefit, it is noted but not addressed here because Congress appears to be firmly committed, at least in the near future, to funding the Human Genome Project at a significant level.

It is clear, however, that efforts to map the human genome and to identify and reduce the harms that the resulting knowledge may create will take place in the context of pervasive racism, ethnic stereotyping, and economic disparities in both health status and access to health care services. Moreover, designing policies to take account of these problems will be difficult. It will potentially require systemic changes in the way health care is delivered in the United States. The needed reforms may even slow the pace of scientific advancement. Yet, I believe, we should not proceed with the gene mapping effort unless we simultaneously commit ourselves to exploring and resolving the issues of fairness and justice that will inevitably be posed by the use of genetic information in a society riddled with race, ethnic, and class divisions. It would be tragic and unjust if the gene mapping endeavor resulted in benefitting only a few and, at the same time, exacerbated existing disparities between groups in our population.

Historical Uses of Genetic Information

The question of whether racial and ethnic groups and the poor can expect fair treatment is of particular concern in conjunction with the generation and utilization of genetic information. In the past, one of the ways in which this society has expressed its preference for homogeneity and rationalized the exclusion of certain groups is by focusing on genetic explanations for perceived difference among groups.[10] This has been particularly true with reference to the status of African-Americans, where it has been all too easy to connect genetic information (often inaccurately portrayed) with racial stereotypes to justify differential treatment of individuals and groups.

The Centrality of Race

Although the concept of race has no scientific or analytical meaning,[11] it has been used historically to describe human differences. It is a socially constructed concept in which differences in phenotypic characteristics of individuals are given prominence; thus, "race in the biological sense has no biological consequences, but what people believe about race has very profound social consequences."[12] In the United States, race is a term used to depict differences between peoples of European and African descent typically in ways that inflict social and economic disadvantage on the latter. As a result, the concept of race is central to an understanding of American history.

From the beginning, it was necessary to demonstrate the superiority of white persons in order to justify the subjugation and enslavement of persons of African descent. Even though the Emancipation Proclamation was issued more than 125 years ago, the matter of race has not been laid to rest. Problems associated with racial inequality are persistent and pervasive. Despite their growing numbers in the population, persons of African descent lack power and thus suffer social and economic disability as a result of being denied access to privileges and opportunities available to others. Negative stereotypes of African-Americans and other minorities persist.[13] Former President Ronald Reagan's use of the phrase "welfare queen" and the power of the Willie Horton image in the 1988 presidential campaign are recent evidence of the power of racial stereotypes. A 1990 National Opinion Research Center survey concluded that whites persist in holding negative and false images of blacks and other minorities as more likely to prefer welfare to work, more likely to be violent, more likely to be lazy, and less likely to be intelligent than white Americans.[14] As the Edsalls observe, "Race is no longer a straight forward, morally unambiguous force in American politics; instead considerations of race are now deeply embedded in the strategy and tactics of politics, in competing concepts of the function and responsibility of government, and in each voter's conceptual structure of moral and partisan identity."[15]

The continued presence of strongly entrenched racial and ethnic stereotypes that permeates all aspects of the nation's life raises questions about whether major social institutions can be trusted to operate fairly where they interact with people of color. This question extends to medicine and medical institutions as well. The use of African-American men in a study of untreated syphilis in Macon County, Alabama, is one example of past abuse which raises questions about whether vulnerable groups will be fairly treated.[16] In another example, young African-American women were coerced into accepting sterilization when confronted with threats that their welfare benefits would be terminated if they refused.[17] The eugenics movement and the sickle cell programs in the early 1970s highlight specific concerns about the role medicine and medical institutions have historically played in the utilization of genetic information.

The Eugenics Movement

The best-known social connection between genetics and stereotypic depictions of human beings is the eugenics movement, which was particularly popular in this nation in the early decades of the twentieth century. Eugenics, the study of, or belief in, the improvement of the human species, and the movement that it inspired have been well documented.[18] Although eugenic discourse displayed a positive orientation by advocating reproduction for the biologically and socially fit, in the United States the focus was primarily on preventing the transmission of deleterious genes. As a result, eugenic discourse became entangled with race,[19] mental disability, and other presumed determinants of parental unfitness. Consequently, eugenic policies resulted in the development of institutions to care for the biologically and socially unfit and restrictive immigration policies. The best-known consequence of these policies was the coerced sterilizations of massive numbers of immigrants, the poor, and institutionalized persons. These state-sanctioned sterilizations were upheld by the Supreme Court in *Buck v. Bell*,[20] a 1927 case that, while seriously undermined, has never been overruled.

Sorenson notes that in the period from the beginning of the twentieth century to the 1930s, "the eugenics movement was a social force in which law played the major regulatory role. Built primarily on voluntary organizations, the movement relied on the mechanisms of law including court decisions, administrative regulations and legislative enactments to achieve its goals."[21]

Although I do not discount the possibility of governmental support for eugenic policies in the future, I believe that the real danger to the interests of racial and ethnic minorities and the poor are likely to come from another direction.[22] People can be harmed and coerced by a web of practices that arise outside of government with consequences as devastating as those associated with governmentally initiated eugenic practices.

Since the 1940s, applied genetics has increasingly become an integral part of clinical medicine. Therefore, knowledge about the values and norms of health care professionals is important to an understanding of the current uses of genetic information. Yet, the role of health care professionals in eugenically inspired theories and policies has only recently been explored, leaving much research in this area yet to be done.[23] Based on this history, my fear is that private practices, more subtle than legal actions—particularly in medicine—will result in stigma, coercion, and discrimination particularly with respect to people of color. This fear gains support from an examination of the practices of sickle cell screening programs.

Screening for Sickle Cell Trait and Disease

Sickle cell anemia is a genetic disorder of varying severity that occurs with greatest frequency in the United States among people of African-American descent. In the early 1970s, sickle cell anemia was commonly regarded as a neglected health problem.[24] As a result, the African-American community and its leaders gave high priority to resolving health problems associated with this condition. Only later did these leaders come to understand that measures intended to call attention to the poor status of health care provided African-American communities would be used to stereotype and disadvantage the very people they sought to help.

Sickle cell screening programs developed as a means of identifying individuals who carried the recessive gene for sickle cell. They were instituted despite the fact that at the time, there were no ways to detect sickle cell trait or disease in utero and there were no means of ameliorating or curing sickle cell disease after birth. The only way to avoid giving birth to a child with sickle cell anemia was not to have children.

Sickle cell screening programs were often mandated by state legislatures, in contrast to voluntary screening programs for Tay-Sachs disease, which primarily affects Jews.[25] Only after the passage of the National Sickle Cell Anemia Control Act in 1972, which required that federal funds be used in voluntary programs, was the movement toward mandatory laws halted.[26] These programs developed in an ad hoc fashion with disastrous consequences for the African-American community.[27] Inadequate provisions for confidentiality led to stigmatization and discrimination in employment and insurance. As a result of inadequate education and counseling, the difference between sickle cell trait and sickle cell disease was poorly understood by those affected and the general community.[28]

The Committee for Study of Inborn Errors of Metabolism concluded that "[s]ome of the early practices of sickle cell screening programs are lessons in what not to do; we may look forward to the useful employment of the wisdom of the veterans of these projects to provide guidance to initiators of future screening programs for other conditions."[29] I hope that the optimism of the Committee is warranted. I believe that our experience with sickle cell screening and eugenic policies teaches that in the United States, when genetics and genetic disease become entangled with race and its accompanying stereotypes, racial and ethnic minorities will suffer.

The Future Potential for Race-Based Discrimination

It is particularly important that we understand lessons from the past because current and future gene mapping efforts may adversely affect minority groups and the poor in at least two ways. First, the genetic information produced will be complex and difficult to interpret. As a result, it will be highly amenable to misinterpretation and abuse. It could be used to legitimate existing racial and class disparities on the grounds that race and class are linked to genetically based characteristics. In other words, we risk increasing social intolerance for differences among human beings. The "biologic underclass"[30] may turn out to be composed of minority group members and the poor. Second, the promise of medical benefit may not be achieved. Medical institutions that direct the distribution of benefits that potentially will flow from gene mapping efforts may not be established in ways that maximize potential benefit for minorities and the poor.

Current gene mapping efforts may uncover fairly easily understood connections among a single gene or chromosome, a disease, and an ethnic group, such as the high correlation between sickle cell anemia and African-Americans. Indeed, in recent years many diseases and conditions, such as cystic fibrosis, polycystic kidney disease, and neurofibromatosis, have been mapped and linked with specific chromosomes.[31] It is also likely that gene mapping efforts will help develop the capacity to predict a person's susceptibility to diseases such as cancer, mental disease, and heart disease.[32] Such associations between genetic information and disease may result in discrimination against all persons who have the genetic susceptibility.

If one of the diseases occurs more frequently among particular minority populations, two additional negative consequences could follow. First, members of the minority group who do not have the susceptibility at issue might nevertheless be discriminated against because of the correlation between disease susceptibility and minority status. This is not a far-fetched possibility. Past discrimination against racial and ethnic minorities tends to support it. In addition, existing discrimination against gay men who are not infected with the human immunodeficiency virus (HIV) virus, but who are members of a vulnerable group associated with HIV infection, suggests that such stigmatization is likely to have a broad sweep. Second, because of its higher frequency among the members of an identifiable minority group, the disease might be given lower priority in terms of research for treatment or cure. Certain populations have historically been ignored when health priorities have been established. Indeed, George Lundberg, Editor of the *Journal of the American Medical Association (JAMA)* notes that "there are many reasons for this [lack of access by many Americans especially blacks and Hispanics to basic medical care] not the least of which is long-standing, systematic, institutionalized racial discrimination."[33] Only recently has there been recognition of the fact that it might be necessary to target special at-risk populations in order to narrow health disparities between these groups and the general population.[34]

The possibility that correlations between genetic susceptibility to disease and group membership will harm racial and ethnic minorities is heightened by the epidemiological practice in the United States of focusing on racial differences alone, rather than on race *and* class, in measuring morbidity and mortality.[35] This practice has the potential of magnifying the association between genetics and minority status. If we look only or primarily at race or ethnicity, it may appear that minority status alone is responsible for different health outcomes.

Avoiding the adverse consequences of expanding genetic knowledge for racial and ethnic minorities and the poor will not be easy. Understanding how or whether minority status correlates with disease is a very complicated problem. There has been some reluctance to call attention to the connection between minority status and disease because of past experiences with eugenics as well as increased understanding of the role environment plays in the development of disease. However, recent research has brought growing attention to the possibility that race and ethnic group membership may be correlated with disease.[36] For example, there is reason to believe that the incidence of severe kidney disease is higher among blacks than whites.[37] There is also some indication (not yet adequately confirmed) that early intervention with zidovudine (AZT) may harm African-Americans with HIV infection while it helps prolong life in white HIV victims.[38]

It is frequently not clear why a correlation exists between a disease and a specific group of people. It will be difficult to discover how culture, socioeconomic status, and environment mediate associations between genetics and race, but these complex interactions may better explain the incidence of many diseases that are in part hereditary.[39] In some cases, a predisposition to a disease may cause health problems. But this predisposition may also be aggravated by environmental factors such as diet. In other cases, it may be impossible to determine the impact of various "causal" factors. For example, a recent Centers for Disease Control study concluded that about one-third of the difference in premature death rates between blacks and whites (the premature death rate for blacks is two to two and one-half times the rate for whites) could be explained by class considerations. Another third of the difference in premature death rates was due to differences in risk factors such as smoking. A full 31 percent of the difference could not be accounted for by any known or measurable variable. The difference could be due to environment, life-style, access to health care, and perhaps heredity.[40]

Great care must be taken in drawing conclusions about the causes of disease based on statistical correlations. For example, a recent study found a correlation between skin color and blood pressure in low socioeconomic groups. The authors suggest that their findings may be due to the lesser ability of such groups to deal with the psychosocial stress associated with darker skin color. However, they caution that the findings are also consistent with an interaction between an environmental factor associated with class and a genetic susceptibility having a higher prevalence among persons with darker skin color.[41]

Understanding the interrelationships of various causal factors in the incidence of disease is exceedingly difficult. As sociologist Troy Duster observes, "[w]hen is a disorder a genetic disorder, and when is it something else?"[42] Trying to understand the role of ethnicity and culture in the incidence of disease without increasing racial stereotyping will be even more difficult. There will be great temptation to rely on simple explanations for group differences. This tendency to simplify is clearly evident in the enduring debates in this nation about the connection between genetics, intelligence, and achievement and continuing efforts to explain behavior in terms of genetics. The current twist to this argument is to urge that efforts to culturally diversify are wrong-headed and discriminatory to whites because of "differing average endowments of people in the two races."[43] As Duster points out, "Once you found Tay-Sachs in the Jews, sickle cell anemia in blacks, beta-thalassemia in Mediterraneans, cystic fibrosis in northern Europeans, you suddenly had a folk logic emerging in both the scientific community and those who knew about these developments. If sickle cell anemia is race-specific, then maybe criminality or intelligence is."[44] The danger to racial and ethnic minorities and the poor from current gene mapping efforts is obvious: the danger is that greater attention will be paid to genetic explanations than to more complex explanations for differences to the detriment of vulnerable and disadvantaged groups.

The Provision of Clinical Genetic Services

If the promise of the Genome Project is realized, distribution of the ensuing medical benefits will be channeled through institutions charged with the responsibility of providing clinical genetic services. Therefore, these institutions should be established and operated so that benefits are fairly distributed and vulnerable groups in society are not disadvantaged. There are several areas of concern that should be central to policy-making decisions: access to services, counseling, and follow-up services. Pilot projects for the provision of genetic services in connection with the introduction of cystic fibrosis screening tests into clinical practice are being developed. Although these projects may be able to identify obstacles to adequate counseling for poor people, they are not likely to focus on obstacles that members of ethnic or racial minority groups will encounter, because cystic fibrosis is a disease that primarily affects whites with northern European ancestry.

Access to Services

A significant issue in the provision of clinical genetic services is the issue of access, or, more precisely, lack of access or unequal access.[45] With respect to

African-Americans, this should come as no surprise in view of the fact that until 1964, health care services for African-Americans were provided for a significant portion of that group on a segregated basis.[46] Although data are scarce, the information we have suggests that access to voluntary genetic screening for conditions that may affect a variety of groups is limited primarily to white, middle-class individuals. For example, women who make use of prenatal diagnosis are disproportionally white, well educated, and financially secure.[47] However, the available data do not support the view that different cultural beliefs concerning disease and abortion account for the paucity of African-Americans who avail themselves of screening technologies.[48] Given the fact that between one-third and one-fourth of pregnant women in the United States do not receive early and continuous prenatal care and that minority women are significantly less likely to receive prenatal care,[49] it should come as no surprise that minority and poor women are unlikely to use genetic screening during their reproductive years. This conclusion is supported by evidence that utilization of genetic services may be related to physician referrals.[50]

The utilization of clinical genetic services programs by poor and minority women seems low despite the fact that all state Medicaid programs surveyed in 1988 did cover amniocentesis, and many programs had expanded coverage to include more recently developed prenatal diagnostic procedures for eligible women.[51] Although Medicaid provides insurance for the poorest families, it does not cover all individuals or families below the poverty line nationwide. In fact, approximately 33 million Americans are not covered by any private or public program. These 33 million Americans are disproportionately low income, black or Hispanic, and young. In those instances where poor persons are covered under Medicaid, only 13 state programs paid for abortion after diagnosis of an anomalous fetus.[52] Moreover, a recent Supreme Court decision holds that recipients of federal funding for family planning services may not inform pregnant women about the option of abortion or refer them to an abortion provider.[53] Clearly, the issues of access and utilization of clinical genetic services programs are matters that deserve further investigation.

Counseling

Even if access to and utilization of genetic screening programs are improved, members of ethnic or minority groups and the poor face significant obstacles in benefitting from genetic counseling and other genetic services. Genetic screening programs follow an ethical norm of material disclosure and nondirective counseling. The goal is that fully informed clients will be empowered to make their own decisions about personal well-being and reproductive options in a manner consistent with their personal values and preferences. Making this norm a reality for racial and ethnic minorities and the poor will be difficult since, unfortunately, communication in cross-cultural contexts is often problematic and not well understood.

In order for genetic information to be beneficial to racial and ethnic minorities and the poor, it is important that genetic clinical services programs understand the concept and significance of human differences and how these differences critically interact with the utilization of genetic information. Terms such as "black," "Hispanic," and "Native American" are imprecise because they group together individuals who in fact affiliate with many different cultures, minority groups, and social classes. In the counseling context, therefore, use of the term "ethnic group" or "ethnicity" is preferable. A number of models have been devised to explain what is meant by ethnic group or ethnicity.[54] Although these models differ in significant ways, they share common elements. "[M]embers of ethnic groups have a sense of a shared past and similar origins . . . believe themselves to be distinctive from others in some significant way . . . [and] ethnicity is most important at those times when members of differing groups are in contact."[55] Ethnicity does not have a biological or genetic foundation.[56] Rather, ethnicity is a classification that refers to a shared social and cultural heritage. Thus, in the sense that classifications are used here, Jews are not a biological race, but instead an ethnic group.

The preference for cultural homogeneity in our society has disposed social institutions, including those offering clinical genetic services, to assume that everyone shares similar values. Social service counseling is based on the values and interests of the dominant group or profession: white, middle-class, and English-speaking. Moreover, little attention has been paid to the nature and significance of informed consent in a variety of cultural contexts. Hahn postulates that two critical aspects of informed consent as understood by health care professionals—competence and rationality—may in fact have culturally specific meanings.[57] Clients can only make truly autonomous choices when information is shared with them in culturally acceptable ways—ways in which their sense of participation and power is heightened. Yet, there is often no awareness on the part of counseling professionals that they bring a distinct cultural perspective to counseling sessions. Bridging cultural differences must begin with training that helps medical professionals to see "themselves as cultural beings whose work profoundly affects and is affected by their own 'selves'."[58]

Unfortunately, rather than seeing themselves as cultural beings whose culture might be different from that of the clients, many counselors are likely to perceive differences between themselves and clients as evidence of the clients' inability to measure up to cultural norms. Yet, ethnicity is an important source of individual and social identity. Members of ethnic groups and the poor may have attitudes, cultural beliefs, and traditions concerning health and health care that are intertwined with cherished group values but that differ significantly from beliefs and values held by the health professionals providing genetic services.[59] They may speak another language. In addition, members of these groups may have a generalized distrust of institutions, professionals, and bureaucracies that is foreign to health care providers. Reform is needed to increase "understanding of the relevance of patients' cultural contexts in the causation of disease, in processes of therapy, and in the formation of understandings, values, and ways of communication that guide the response and conduct of patients and others in health and illness."[60]

The issue of culturally and ethnically appropriate counseling is further complicated, however, because individual clients may not necessarily fit into traditional stereotypes. For example, a person who is a member of an ethnic group may be more influenced by class norms than ethnic status. Thus, it should not be assumed that increasing the numbers of ethnic and minority group members as counselors will, in and of itself, assure culturally appropriate counseling, although such measures will, no doubt, help to ease cross-cultural communication. Maintaining a balance between sensitivity to cultural variation and avoidance of sterotyping is not easy for any counselor regardless of his or her own ethnic background.

The difficulties of providing nondirective counseling to large numbers of culturally diverse and poor clients may prove so severe that the norm of enhancing individual autonomy may itself be undermined. There are indications of this breakdown in discussions about HIV infection and reproduction. In the United States, HIV infection in women is primarily a problem for low-income African-American and Hispanic women. Their behavior has challenged implicit norms about reproduction and parenting. Some HIV-infected women continue to have sexual intercourse without using contraceptives or fail to use barrier methods of contraception. Some pregnant women do not abort after learning that they are HIV-positive.[61] The failure to appreciate different cultural values[62] leads some to consider directive counseling in some circumstances.[63] It may turn out that the perceived success of full disclosure and nondirective counseling in clinical genetic programs is due to the fact that clients shared the values of their counselors; therefore, nondirective counseling may not be an achievable goal for future counseling programs.

Moreover, it is unclear whether nondirective counseling norms and the value such counseling places on individual choice can or should be maintained in the face of population genetic screening that emphasizes public health goals, most notably the reduction of the number of children born with severe impairments. Some fear that such population screening will constitute a revival of coerced reproductive "choices." For racial and ethnic minorities and the poor, the fear is that they will be disproportionately the targets of such coercive practices.

As pointed out earlier, coerced reproductive decisions have been the basis for social and public health policies in the past. More recently, the availability of Norplant (an implantable long-term contraceptive) has raised anew the question about whether poor women should have their reproductive options restricted. A *Philadelphia Inquirer* editorial suggested providing Norplant to women on welfare.[64] A judge in California required use of Norplant as a condition of probation in the case of a woman pleading guilty to child abuse.[65] Legislation was introduced in Kansas that would offer money to any woman with children on welfare who agreed to use Norplant.[66] A recent *Los Angeles Times* poll found that 61 percent of the persons polled approved of making Norplant mandatory for drug-abusing women of childbearing age.[67] These responses to the introduction of Norplant into clinical medicine raise the possibility that the emphasis on autonomy and individual choice, particularly in the reproductive context, may be compromised in situations involving clients who are subject to pejorative racial and ethnic stereotypes and who, in fact, may have different norms and values regarding parenting and reproduction.

Provision of Follow-up Services

Even if one assumes that counseling can be made culturally appropriate for individuals and that the values of individualism and choice will be maintained or, if altered, will be applied fairly to all, there is no reason to believe that racial and ethnic minorities and the poor will receive the level of health services necessary to prevent, ameliorate, or cure the genetic conditions they have been counseled about. Disparities in health status and access to health care services already exist between these groups and others in the population. For example, a recently published study evaluated mortality for twelve diseases for which death is avoidable given timely and appropriate medical intervention. For these twelve diseases, the mortality rate of blacks was *four and one-half times that of whites.*[68] Without changes in our system of health care financing and access to health care delivery, there is no reason to believe that members of socially disadvantaged groups will benefit from the information they receive. Moreover, focusing on health care services is only a part of the problem. Social and economic factors, such as inadequate housing and lack of employment, are also significant factors in the incidence of disease, which, like many other problems suffered primarily by disadvantaged groups, receive low priority, particularly in times of scarce resources and public disillusionment with social welfare programs.

Conclusion

In the face of obstacles to securing quality health care, it is unlikely that racial and ethnic minorities and the poor as groups will benefit from the Genome Project without substantial changes in our existing health care system. Moreover, racial and ethnic communities and their leaders are unlikely to have the energy to address problems raised by emerging scientific developments when they already face insurmountable obstacles to rectifying present-day problems. From the perspective of minority group leaders, it might be preferable to advocate a reordering of priorities in light of the potential dangers the Genome Project poses to the interests of their constituents. Allocating limited resources to ameliorate economic, social, and environmental conditions that influence health status would yield more immediate and enduring health benefits than focusing on genetic contributions to disease prevention.

Since the pressing interests of minority group leaders may of necessity lead them away from involvement with the Genome Project, efforts to increase the likelihood that beneficial outcomes from new genetic information will be fairly distributed must come primarily from those involved with gene mapping. Some, I am sure, will assert that those who are charged with advancing science cannot be held responsible for the misapplication of the information produced by their efforts or the existing social ills that are prevalent in the society. Yet, it is society that supports their work. Society can reasonably require that those with intimate knowledge of science and its social implications participate in efforts to benefit all. The creation of ELSI is evidence of a broad consensus that the advancement of knowledge must be simultaneously accompanied by efforts to mitigate or prevent misapplication and abuse of knowledge gained. Attempts to remedy inequalities in health care delivery and simultaneously to avoid racism and stereotyping as a consequence of more and improved genetic information will be difficult. However, those who are charged with considering the ethical, legal, and social implications of gene mapping must, at a minimum, take account of the social and policy implications of differences among human beings in trying to achieve a favorable balance of risk to benefit in policy decisions about the use of new genetic information. As an integral part of that balance, they must seek to ensure that benefits, should they materialize, are fairly distributed to all. I believe that the goal can only be achieved by resolving existing health care problems as well as anticipating new ones.

ACKNOWLEDGMENTS

I am deeply indebted to many friends and colleagues, including Steve Goldberg, Lani Guinier, Emma Coleman Jordan, Gerry Spann, Robert Suggs, Zoe Ulshen, and Roger Wilkins, who read and critiqued earlier drafts of this chapter.

NOTES

1. V. A. McKusick, "Mapping and Sequencing the Human Gene," *New England Journal of Medicine,* 320:910–15, 1989, and Chapter 2, this volume.
2. National Commission for the Protection of Human Subjects of Biomedical and Behavioral Research, *The Belmont Report: Ethical Principles and Guidelines for the Protection of Human Subjects of Research,* Government Printing Office (DHEW Publication No. (OS) 78-0012), Washington, D.C., 1978, p. 7.
3. T. L. Beauchamp and J. F. Childress, *Principles of Bioethics,* 3rd ed., Oxford University Press, New York, 1989, p. 258.
4. This question is also relevant to gender, which, regrettably, I cannot pursue in this chapter.
5. R. A. Shweder, "Dangerous Thoughts. . . ." (book review of C. N. Degler, *In Search of Human Nature: The Decline and Revival of Darwinism in American Social Thought,* Oxford University Press, New York), *New York Times Book Review,* March 17, 1991, p. 1.
6. Committee for the Study of Inborn Errors of Metabolism, *Genetic Screening Programs, Principles, and Research,* National Academy of Sciences, Washington, D.C., 1975.
7. R. J. Herrnstein, "Still an American Dilemma," *The Public Interest* 98:3–17, 1990; and P. Selvin, "The Raging Bull of Berkeley," *Science* 251:368–71, 1991.
8. U.S. Department of Health and Human Services, Public Health Service, *Healthy People 2000,* Government Printing Office (DHHS Publication No. [PHS] 91-50212), Washington, D.C., 1991.
9. G. J. Annas, "Mapping the Human Genome and the Meaning of Monster Mythology," *Emory Law Journal* 39(3):629–64, 1990.
10. J. W. Green, *Cultural Awareness in the Human Services,* Prentice-Hall, Englewood Cliffs, NJ, 1982, p. 59.
11. Ibid.
12. D. R. Atkinson, G. Morten, and D. W. Sue, *Counseling American Minorities: A Cross-Cultural Perspective,* 3rd ed., Wm. C. Brown Publishers, Dubuque, IA, 1989, p. 4.

13. J. Kirschenman and K. M. Neckerman, "We'd Love to Hire Them, But . . .": The Meaning of Race for Employers. In *The Urban Underclass,* edited by C. Jencks and P. E. Peterson, The Brookings Institution, Washington, D.C., 1991, pp. 203–34.

14. L. Duke, "Whites' Racial Stereotypes Persist, Most Retain Negative Beliefs about Minorities, Survey Finds," *Washington Post,* January 9, 1991, p. A1.

15. T. B. Edsall and M. D. Edsall, "When the Official Subject Is Presidential Politics, Taxes, Welfare Crime, Rights, or Values . . . The Real Subject Is Race," *Atlantic,* May 1991, pp. 53–86.

16. A. M. Brandt, "Racism and Research: The Case of the Tuskegee Syphilis Study," In *The Social World,* edited by I. Robertson, Worth, New York, 1981, pp. 186–195. J. Jones *Bad Blood: The Tuskegee Syphilis Experiment,* Free Press, New York, 1981.

17. Relf v. Weinberger, 565 F.2d 722 (D.C. Ct. App., 1977).

18. D. J. Kevles, *In the Name of Eugenics: Genetics and the Uses of Human Heredity,* Alfred A. Knopf, New York, 1985.

19. P. R. Reilly, "Eugenic Sterilization in the United States," In *Genetics and the Law III,* edited by G. J. Annas, Plenum Press, New York, 1984, pp. 227–41.

20. *Buck* v. *Bell,* 274 U.S. 200 (1927).

21. J. R. Sorenson, "From Social Movement to Clinical Medicine—The Role of Law and the Medical Profession in Regulating Applied Human Genetics," In *Genetics and the Law,* edited by A. Milunsky and G. Annas, Plenum Press, New York 1975, pp. 467–85.

22. T. Duster, *Back Door to Eugenics.* Routledge, Chapman and Hall, New York, 1990.

23. See note 9.

24. President's Commission for the Study of Ethical Problems in Medicine and Biomedical and Behavioral Research, *Screening and Counseling for Genetic Conditions: A Report on the Ethical, Social, and Legal Implications of Genetic Screening, Counseling, and Education Programs,* Government Printing Office, Washington, D.C., 1983, p. 21.

25. See note 22, pp. 45–50.

26. See note 24, p. 21.

27. J. Bowman and R. F. Murray, Jr., *Genetic Variation and Disorders in Peoples of African Origin.* Johns Hopkins University Press, Baltimore, 1990, pp. 365–6.

28. See note 6, pp. 116–33.

29. Ibid, p. 127.

30. D. Nelkin and L. Tancredi, *Dangerous Diagnostics: The Social Power of Biological Information,* Basic Books, New York, 1989.

31. See note 1.

32. Ibid.

33. G. Lundberg, "National Health Care Reform: An Aura of Inevitability Is upon Us, *Journal of the American Medical Association* 265:2566–7, 1991.

34. See note 8.

35. V. Navarro, "Race or Class versus Race and Class: Mortality Differentials in the United States," *Lancet* 336:1238–40, 1990.

36. W. E. Leary, Uneasy Doctors Add Race-Consciousness to Diagnostic Tools, *New York Times,* September 25, 1990, p. C1, col. 1, and see note 27.

37. B. L. Kasiske, et al., "The Effect of Race on Access and Outcome in Transplantation," *New England Journal of Medicine* 324(5):302–7, 1991

38. "Race Joins Host of Unanswered Questions on Early HIV Therapy," *Journal of the American Medical Association 256:1065–6, 1991.*

39. C. R. Baquet, et al., "Socioeconomic Factors and Cancer Incidence among Blacks and Whites," *Journal of the National Cancer Institute* 83:551–567, 1991.

40. M. Gladwell, "Poverty Major Cause of High Black Death Rate," *Washington Post,* February 9, 1990, p. A1.

41. M. J. Klag, et al., "The Association of Skin Color with Blood Pressure in U.S. Blacks with Low Socioeconomic Status," *Journal of the American Medical Association* 256:599–640, 1991.

42. See note 22.

43. See note 7, Herrnstein, 1990.

44. See note 7, Selvin, 1991, p. 369.

45. J. C. Fletcher, "Ethics and Human Genetics: A Cross-Cultural Perspective." In *Ethics and Human Genetics: A Cross-Cultural Perspective,* edited by D. C. Wertz and J. C. Fletcher, Springer-Verlag, New York, 1989, p. 463.

46. M. T. Rice, "Black Hospitals: Institutional Impacts on Black Families." In *Black Families: Interdisciplinary Perspectives,* edited by H. E. Cheatham and J. B. Stewart, Transaction Publishers, New Brunswick, NJ, 1990, pp. 49–68.

47. J. C. Fletcher and D. C. Wertz, "Ethics, Law, and Medical Genetics: After the Human Genome Is Mapped," *Emory Law Journal* 39:747–810, 1990. M. M. Adams, et al., "Utilization of Prenatal Genetic Diagnosis in Women 35 Years of Age and Older in the United States, 1977 to 1978," *American Journal of Obstetrics and Gynecology* 139:673–7, 1980. S. J. Sepe, et al., "Genetic Services in the United States," *Journal of the American Medical Association* 248:1733–5, 1983. J. P. Marion, et al., "Acceptance of Amniocentesis by Low-Income Patients in an Urban Hospital," *American Journal of Obstetrics and Gynecology* 1980; 138:11–5, 1980.

48. Ibid, Marion, 1980.

49. S. S. Brown (editor for the Institute of Medicine), *Prenatal Care: Reaching Mothers, Reaching Infants,* National Academy Press, Washington, D.C., 1988.

50. See note 47, Fletcher and Wertz, 1990, p. 761.

51. J. Weiner and B. A. Bernhardt, "A Survey of State Medicaid Policies for Coverage of Abortion and Prenatal Diagnostic Procedures," *American Journal of Public Health* 80:717–20, 1990.

52. Ibid.

53. *Rust* v. *Sullivan,* 111 Sup. Ct. 1759 (1991).

54. F. Barth, ed. *Ethnic Groups and Boundaries: The Social Organization of Cultural Difference,* Little Brown, Boston, 1969, and note 10, Green, 1982, p. 9.

55. See note 10, Green, 1982, p. 9.

56. See note 12, Atkinson et al., 1989, p. 4.

57. R. A. Hahn, "Culture and Informed Consent: An Anthropological Perspective," In *Making Health Care Decisions: The Ethical and Legal Implications of Informed Consent in the Patient-Practitioner Relationship,* President's Commission for the Study of Ethical Problems in Medicine and Biomedical and Behavioral Research, Government Printing Office, Washington, D.C., 1982, p. 59.

58. Ibid.

59. R. F. Murray, "Special Considerations for Minority Participation in Prenatal Diagnosis," *Journal of the American Medical Association* 243:1254–6, 1980.

60. See note 57.

61. P. Selwyn, et al., "Knowledge of HIV Antibody Status and Decisions to Continue or Terminate Pregnancy among Intravenous Drug Users, *Journal of the American Medical Association* 261:3567–71, 1989.

62. C. Levine and N. N. Dubler, "Uncertain Risks and Bitter Realities: The Reproductive Choices of HIV-Infected Women," *Milbank Quarterly* 68:321–52, 1990.

63. J. D. Arras, "AIDS and Reproductive Decisions: Having Children in Fear and Trembling," *Millbank Quarterly* 68:353–82, 1990.

64. "Poverty and Norplant: Can Contraception Reduce the Underclass?" (editorial), *Philadelphia Enquirer,* December 12, 1990, p. 18-A, col. 1.

65. W. Booth, "Judge Orders Birth Control Implant in Defendant," *Washington Post,* January 5, 1991, p. A1, col. 3.

66. T. Lewin, "A Plan to Pay Welfare Mothers for Birth Control," *New York Times,* February 9, 1991, p. 9, col. 4.

67. G. Skelton and D. M. Weintaub, "Most Support Norplant for Teens and Drug Addicts," *Los Angeles Times,* May 27, 1991, p. 1, col. 3.

68. E. Schwartz, et al., "Black/White Comparisons of Deaths Preventable by Medical Intervention: United States and the District of Columbia 1980–1986," *International Journal of Epidemiology* 19:591–8, 1990.

III

The Human Genome Project and the Human Condition

6

Determinism and Reductionism: A Greater Threat Because of the Human Genome Project?

Evelyne Shuster

The leaders of the Human Genome Project have consistently described it in deterministic and reductionist language. For example, James Watson characterized the Project as the search for "ultimate answers to the chemical underpinnings of human existence"[1] and stated that "in large measure, our fate is in our genes."[2] Others have claimed that "knowing the complete human genome we will know what it is to be human."[3]

These hyperbolic statements support a view that genetic knowledge is the ultimate in determinism and hereditarianism.[4] The conceit is that once the structure and function of the genome is understood, once the concepts of genetic code, program, and messages are grasped, it may seem possible to have a gene-based explanation of all phenotypic characteristics, including all aspects of human health, disease, and even behavior. It has already been suggested by the editor of *Science* that the knowledge gained from the Human Genome Project could solve problems of homelessness and crime and "aid the poor, the infirm and the underprivileged."[5] A California appellate court, the first appeals court to rule on this issue, has also relied solely on maternal genes to deny a gestational mother parental rights over the child she bore but to whom she was genetically unrelated.[6]

Genetics is no longer a domain of pure speculation. Clinical geneticists now work to discover genetic defects, diagnose or predict genetically influenced diseases, screen, counsel, and assist in human reproduction. A perception that human genetics is essentially deterministic and reductionist could lead to the misapplication of genetic information and foster socially dangerous ideologies. Just as the Nazi physicians enthusiastically misused genetics to promote and implement their racial hygiene program in the 1930s and 1940s,[7] so too others could misuse the fruits of the Human Genome Project. A *misuse* of genetic information could justify the destruction of all embryos less than "perfect," the de facto creation of a new "biological underclass," and the systematic ostracism of the "genetically unfit." Society could have a powerful genetic tool for controlling individuals through an entire series of labeling and intervention: a *"bio-politics of the population."*[8]

None of this, of course, is necessary. It should be obvious that the attainment of a complete map and sequence of the genome will not provide a solution to human problems. Nor will it explain what makes humans uniquely human. But, the perception has been that what is genetic is unchangeable, and that problems of criminality, behavioral deviation, individual capability, even differences between sex, race, and general intelligence (IQ) can be accounted for solely from within the domain of human genetics.[9] Ultimately, perception is all that matters. If it cannot be persuasively dispelled, the applicability of genetic information in predictive and curative medicine and in practical human affairs will be problematic at best and could be dangerously attractive and destructive of cultural and moral interests.

The Mechanistic Model: The Body-Machine

Modern biologists believe that the description of biological phenomena is far less meaningful than the elucidation of the molecular mechanisms underlying them. They "work with a certainty that the invisible, submicroscopic agents they study can explain, at one essential level, the complexity of life."[10] Biological systems are understood in terms of molecular mechanisms rather than in terms of organisms, organ system, or organs. This research strategy has been spectacularly successful (for example, in diagnosis of genetic defect, artificial reproduction, and genetic engineering). The working assumption is that once the exact sequence of nucleotides contained in the genes has been identified, the one-to-many relation from the molecular to the phenotypic level will have been converted to a one-to-one relation. The view that humans can ultimately be accounted for solely from their molecular structure has been the central fear, and a reason for society's mistrust of the new biology. This is because reducing humans to molecular components, and the body (once cultural) to biochemical reactions, changes the way we think about ourselves as unique individuals, lessens the value of life, and undermines the notions of individual worth, freedom, and responsibility.

To be sure, reductionistic models applied to humans are not new.[11] For example, starting in the seventeenth century with the new physics, the belief was that all of science, including "biology," could be derived from mechanics. In this view, a standard machine could serve as a metaphoric model for the understanding of humans. For instance, Descartes focused on the "animal-machine" and characterized the human body as "an ingenious automaton made by the hands of God."[12] He maintained that if organic functions of animals, in general, and of humans, in particular, were to be known, organisms had to be treated quite literally as machines pure and simple, made up of pulleys, wheels, weights, pistons, and meshing gears. His assumption was that the body-machine would lend itself to a complete and transparent (rational) understanding of humans. This understanding would be the basis for a rational, that is, mechanistic, medicine. For Descartes, such medicine would help free men from "an infinity of maladies, both of the body and of the mind, and possibly of all the infirmities of age."[13] It would also "be a means of rendering men wiser," and thus, of controlling human behaviors.[14]

The Cartesian doctrine of the animal-machine reveals serious problems with a strict mechanistic reductionism. Why should a machine model be more appropriate for an animal than for a rock? Paradoxically, it is because a machine cannot be explained in strict mechanistic terms that it can recommend itself as a model for the understanding of organisms. A machine can only be explained from the outside. We ask what task it has been designed to perform. Based on the answer, we can determine whether the machine is "working" well, whether it is not working even though the wheels might be spinning, and whether it can be restored to a good working condition. Built into the concept of a machine are normative criteria for evaluating its internal and external performance. The four Aristotelian causes—formal, material, efficient and final—reappear in a straightforward philosophical fashion. Only when nature is understood as a model of art, as is explicitly the case with Aristotle and only implicitly with Descartes, can doubts arise about interpreting nature as a whole, including humans, from a strict mechanistic perspective. No wonder, then, that Descartes had to abandon his mechanistic and reductionist view of humans. He ultimately confessed that even though everything in the human body can be explained mechanistically, knowledge of bodily functions alone cannot lead to an understanding of humans.[15]

Humans could not be reduced to machines because the body is united to the mind, "the thinking soul," with which it forms an indivisible unity.[16] The "thinking soul" is what makes the reductionist Cartesian model useless. We may have, according to Descartes, a "clear and distinct idea" of the two substances taken separately: the body as pure extension, the soul as a "thinking thing." But when united, they constitute a biological composite that does not lend itself to a strict mechanistic explanation. A purely mechanistic model only works when applied to the classic example of billiard balls colliding and rebounding in accordance with the Newtonian or Cartesian laws, when the player's intents are waived. But when applied to a machine, a clock for instance, the model has a normative and teleological value that cannot be accounted for by mechanistic laws.[17] "Designed for a purpose, a machine only serves that for which it has been built. A strict mechanistic model could not account for both finality and determinism in an organism."[18] It is therefore only through a conflation of meanings and a confusion of concepts that a pure mechanistic model appears to teach us anything about humans.

Efforts at reducing humans to machines failed to provide a useful model, even during the classical era. This suggests that as we move from simple entities (the balls colliding and rebounding) to increasingly complex organisms, the problems of determinism and predictability mount. The inability to establish a causal one-to-one relation between entities in interaction becomes overwhelming as these entities increase in complexity. As the level of knowledge changes from organic structures visible to the naked eye to molecules, and then atoms, the complexity of an organism becomes so great that a science based on deterministic causality is impossible.

The Anatomical Model: Vesalius Revisited

Descartes was determined to abolish Aristotelian teleology from the study of nature and humans. The mechanistic world view that committed him to account for all "vital phenomena" solely in mechanistic and physical terms (efficient causality) would have been inconceivable to Vesalius. This is because Vesalius favored a human-centered organized world. His preference rested on indemonstrable (though plausible) considerations of simplicity, order, and harmony. No wonder Victor McKusick has employed the Vesalian anatomical model, rather than the Cartesian mechanistic model, as the appropriate metaphor for gene mapping and sequencing.[19] To McKusick, it seems safe to compare the clinician–geneticist with the physician–anatomist, and to apply the neo-Vesalian or anatomical model to gene mapping and sequencing. Thus, "the molecular biologist who studies segments of the genome would be like a physician–anatomist who studies organs. He would be in the same position as the nephrologist with his kidney or the cardiologist with the heart. He will have an organ (the gene) that he can biopsy and of which he can analyse disordered structure and function and which he can attempt to repair." Just as the physician–anatomist Andreas Vesalius, in his *De Fabrica Humani Corporis,*[20] could not fail to protect human integrity when dissecting the bodily anatomy, so too, the geneticist who "dissects" genomic anatomy.

Vesalius kept intact humans in their living environment. The Vesalian anatomical description is always accompanied by dynamic and meaningful pictorial representations of humans and observations on the use and functions of the structure described. Never did the artistic drawings of the *De Fabrica* sketch humans as mere anatomical objects to be dissected and passively observed by the physician–anatomist. The Vesalian man stands erect, a posture Aristotle has characterized as essentially human: "Man stands erect because his essence is divine."[21] The old Aristotelian order of nature clearly reappears. This is because the standing position is in harmony with the hierarchical order of the universe that places humans at the top of the "Chain of Being." For Aristotle, no other animals have quite the same posture because they do not have the faculty of reason.

Georges Canguilhem observed[22] that the Vesalian skeletons, with their chests open, stand erect, turning their painful faces toward a civilized and humanized world. The ruins, the bridges, the palaces, the towers and the steeples, the whole Paduan landscape in the background of the drawings reflect human creativity. The Vesalian man lives in a cultural world for which he bears full responsibility. He is simultaneously the creator of values, the artist, the engineer, and the working man. He remains at the center of the world. Never has he become a "passive object" of scientific research. Never does he completely submit to the scientist, the naturalist, the anatomist–physician, or the clinical observer. His movements are deliberate and full of defiance and spontaneity.

The Vesalian anatomical portraits contrast sharply with the human gene maps that the molecular biologist uses to represent life processes. If these illustrations continue to speak of the living being, they do so at such a high level of abstraction that the dynamics and purposefulness of life are reduced to molecular reactions.[23] Biology can no longer be a pictorial representation of humans moralizing over death, enlightened by what they see, and impressed with their own accomplishments (Vesalius). Nor can it be a technical description of humans as simple machines indifferent to themselves and their environments (Descartes).

What seems to characterize the gigantic shift in perspective from the macro to the micro level of enquiry is its aggressive, simplifying determinism and reductionism: "culture is reduced to biology; biology, to the laws of physics and chemistry at the molecular level; mind, to matter; behavior, to genes;

organism, to program; the origin of species, to macromolecules; life, to reproduction."[24] This is captured in these definitions: "The aim of molecular biology is to find in the structures of macromolecules, interpretations of the fundamentals of life."[25] It is to "interpret the properties of the organism by the structure of its constituent molecules"[26] and to "search for explanations of the behavior of living beings in terms of the molecules that compose them."[27] The question therefore remains: is the understanding of the molecular mechanisms of life the ultimate in deterministic reductionism?

The Genotype: A Self-Referential Model

The molecular biologist describes life and evolution in terms of information, messages, and code. The DNA molecule constitutes a "text," a "blueprint," or a "program" by which the "chemical factory" that is the living cell or organism constructs and operates itself. Replications are "reading," and mutations "misreadings," "misspellings," or "copying errors."[28] The computer has become the appropriate model for understanding organisms, and has provided a new language for biologists and a set of rules by which life is understood. A new way of thinking emerges. Organisms are assimilated to information systems that "absorb and store information, change their behaviors as a result of that information and have special organs for detecting, sorting, and organizing this information."[29] Thus, "the organism becomes the realization of a program prescribed by its heredity. . . . Reproduction [of its constituent molecules] represents both the beginning and the end, the cause and the aim."[30]

The metaphors of "codes" and "machinery" can now account for both finality and determinism in an organism, that is, precisely what a strict mechanistic model of humans was ill equipped to handle. As François Jacob pointed out: "[w]ith the possibility of carrying out mechanically a series of operations laid down in a programme, the old problem of the relations between animal and machine was posed in new terms. . . . Animal and machine, each system becomes a model for the other."[31] In this sense, "modern biology belongs to the new age of mechanism,"[32] which readily encourages social policies in support of genetic engineering and the patenting of engineered forms of life, and thus, a reductionist view of humans as "genetic products."

However, Howard Kaye has observed that the "reconceptualization" of life in terms of code, programs, and cybernetic systems is more than a successful research strategy.[33] It specifically represents a world view that defines nature as a system of storage and transmission of information. This representation is therefore no more objective than are any other theoretical models in science: "The reduction of all of biology, all of the behavior characteristics and 'fundamentals of living things' to molecular mechanisms of life betrays a metaphysical ambition to demonstrate that organisms really *are* machines, and that all of life can be accounted for in this way."[34] The world view, more than nature itself or a specific subject matter, has guided scientists' social and ethical speculations and their sensational public pronouncements about the Human Genome Project. Other scientists with a different conception of nature who have worked on the same biological level have had different results. For example, to Barbara McClintock, the genome is

> [a] highly sensitive organ of the cell that monitors genomic activities and corrects common errors, senses unusual and unexpected events, and responds to them often by restructuring the genome. We know about the components of genomes that could be made available for such restructuring. We know nothing, however, about how the cell senses danger and instigates responses to it that often are truly remarkable.[35]

Her appreciation of the genome is far removed from the kind of reductionistic determinism that James Watson articulated in his pronouncement on the overarching significance of the Human Genome Project.

The leaders of the Human Genome Project have thus created, through their world views, a "paradigm shift" in genetics and an aggressive, simplifying, reductionistic perception of genetic knowledge and of humans. Their immediate success in research strategy has enabled them to pass persuasively from science to social implications and to express powerfully their views in the form of reductionist and deterministic generalization in advance of experimental evidence.[36]

The computer analogy may help us better to understand the transmission of information once a program has been established. For example, phenylketonuria (PKU) demonstrates that the conceptualization of life as transmission of information has helped reframe the concept of disease as qualitative "errors" in genetic instruction, rather than quantitative dysfunctions. As a result, the PKU model has been spectacularly successful in guiding treatment of this inherited metabolic error through a diet begun at birth. But the computer model cannot explain the emergence of a program that the genome itself initiates in response to unanticipated challenges because there is no such thing as a "self-constructing machine." Nor can we yet produce, outside of the fiction of *Star Trek: The Next Generation,* a computer that has any of the fundamental characteristics of human life. We may then continue to speak of a "cell as a machine"[37] that obeys the dictates of a genetic program, and to use the computer analogy for understanding molecular functions. Nonetheless, the emergence of life escapes the model intended to explain it. The reductionist and deterministic argument presupposes that an identical genotype invariably produces an identical phenotype, just as identically programmed computers always perform the same functions. This assumption rests on the principle of deterministic causality according to which the same cause always produces the same effect. If the effects are different, then the causes must be different.

On the other hand, "humans, like any other organisms, have been genetically programmed. But they have been programmed to learn. A variety of possibilities has been offered to them by nature at birth. However, that which is actually realized is being constructed during lifetime."[38] This indicates that a genetic program is not fixed and predetermined. It does not constitute a "life text" from which it could be possible to establish a one-to-one relation between the genes and phenotypic characteristics. For example, scientific studies of twins reared apart continue to affirm the major influence that environment has on the individual, an influence that produces far different people even though their genetic makeup is identical[39]: "A genome may reorganize itself when faced with a difficulty for which it is unprepared. . . . The types of response are not predictable."[40] What the genotype specifies is a range of potentially possible phenotypes, "the entire range of reaction, the whole repertoire, of the variant paths of development that may occur in the carriers of a given genotype in all environments, favorable and unfavorable, natural and artificial. *It is fully known for no genotype*"[41] (emphasis added). In a controlled environment, the gene molecules may appear as structurally crystalline, isomeric elements. They can be identified, analyzed, and reduced to molecular reactions. Logically, at this level there can be no workable criteria for distinguishing permissible from impermissible gene intervention. An argument could be made that "looking at the body through the eyes of the clinician–geneticist"[42] reveals an esoteric world of genomic anatomy where humans have lost all essential characteristics.

Crossing the threshold of the living being down to the molecules, and looking at life at its elementary level, results in a mechanistic world of cytogenetics that can hardly be said to belong to a human. However,

> outside the cell, without the means to carry out the plans, without the apparatus necessary for copying or translating, [the genetic program] remains inert, like a tape outside the tape-recorder. No more than memory of a computer can the memory of heredity act in isolation. Able to function only within the cell, the genetic message can do nothing by itself. It can only guide what is being done. To produce machines from plans, there have to be machines. None of the substances that can be extracted from the cell can re-produce. Only bacterium, the intact cell, can grow and reproduce, because only the cell possesses both the program and the directions for use, the plans and the means of carrying them out.[43]

Thus, considered within itself and in interaction with the milieu, an organism functions as a whole whose "parts" are only "parts" to the observer. In a living environment, the organism interacts with the milieu in an unpredictable fashion of virtually infinite possibilities that eludes calculation. The individual's genetic makeup sets outer limits on the phenotype, the range of reactions in which the characteristics of the developing organism must lie. The range of reactions and the number of potential combinations are so staggering that it is impossible to predict which combination has prompted an orderly morphogenesis of a fertilized ovum into a singular multicellular organism. As an organism increases in complexity, genetic instructions offer new capacities for adaptation, new ways of learning and remembering. Some genetic instructions will be excluded, others will emerge, and still others will be translated into new potentialities. But the very operations of the genes are such that they establish their own limits and range of adaptational capability. The very occurrence of these operations is unpredictable. It has no model.

The inability to determine the way mapped genes will be phenotypically expressed demonstrates the limits of a deterministic and reductionist argument applied to humans. In the last analysis, the very conceptualization of life processes begins and ends with a renunciation of explanation of life itself. Prigogine and Stengers best summarize this view, stating:

> The real lesson to be learned from the principle of complementarity [the double-helical, self-complementary structure of DNA], a lesson that can perhaps be transferred to other fields of knowledge, consists on emphasizing the wealth of reality which overflows any single language, any single logical structure. Each language can express only part of reality. Music, for example, has not been exhausted by any of its realizations, by any style of composition, from Bach to Schönberg.[44]

When Descartes offered a reductionist explanation of humans based on the machine analogy, he was hoping to abolish the unpredictability and finality of life. All that occurs in a machine is fully knowable and predictable. The irony is that this very model made it impossible for Descartes to find a cure for "even the most simple fever."[45]

Conclusion

The paradox is perhaps that the molecules of genes operate through a series of instructions that can be spelled out, but emerge in a way that cannot be fully explained. We may be able to retrospectively determine, from a variety of different approaches, what could have predisposed an individual to behavioral traits and diseases. But no model can enable us to predict how the genes will respond to challenges, interact with the milieu, and restructure themselves to ensure their survival.

The charting of the new world of genomic exploration is not reducible to the charting of the human genome, even if this is where it begins. The leaders of the Human Genome Project themselves have begun backing away from their earlier metaphors and hyperboles, which set the stage for the grand (and scary) deterministic and reductionist design. They are now speaking of "building infrastructure"[46] for science and understanding the composition of the genomic anatomy. The infrastructure metaphor is probably more appropriate and sensible than the previous ones. It also provides a more accurate and realistic model for the Genome Project. If the Human Genome Project is the study of the genomic infrastructure, then we may better understand its purpose and priorities. Because humans are becoming subject to genetic study, it seems logical to have in place an infrastructure that allows for systematic, cost-effective research. But research strategy does not warrant acquainting the public with what scientists believe are the social, ethical, and legal implications of their scientific achievements and the discipline of science itself.

On the other hand, in a TV-dominated age, public perceptions are often more important to policy making than reality, and the Human Genome Project will have to deal with its reductionist and deterministic tendencies. We already see in our medical schools the infiltration of molecular and cell biologists into every field and department, to the point where only a few anatomists are left.[47] Performing tests and interpreting their results have become almost more important for the diagnosis and treatment of diseases than interacting with a patient as a person. This is far from the vision of Vesalius and his *Fabrica!*

Obviously, we are only at the beginning of this "molecular revolution." Unquestionably, we will emerge from it with modified views of the genome only to await the emergence of new knowledge that again could bring startling changes in metaphors and concepts. This was clearly understood by François Jacob, for example, who refused to think that the reconceptualization of life in terms of genetic codes, programs, and information provides the essential "truth" about humans. Nor did he give in to the reductionist and deterministic temptations of modern biology. Viewing science and life using the "Russian doll" model, where "the objects of observation fit one inside the other" in an endless pattern, he understood that some level of organization will eventually be discovered and some other set of ruling concepts will be developed.[48] In his words:

> [u]ltimately all organizations, all systems, all hierarchies owe their very possibility of existence to the properties of the atoms described by Clerk Maxwell's

electromagnetic laws. There are perhaps other possible coherences in descriptions. But science is enclosed in its own explanatory system, and cannot escape from it. Today the world is messages, codes and information. Tomorrow what analysis will break down our objects to reconstitute them in a new space? What new Russian doll will emerge?[49]

NOTES

1. J. D. Watson, "The Human Genome Project: Past, Present and Future," *Science* 248:44–8, 1990.
2. Quoted in L. Jaroff, "The Gene Hunt. Scientists Launch a $3 Billion Project to Map the Chromosomes and Decipher the Complete Understanding for Making a Human Being," *Time*, March 20, 1989, pp. 62–71.
3. Walter Gilbert, quoted in R. Lewontin, "The Science of Metamorphoses," *New York Review of Books*, April 1989, p. 18.
4. See V. A. McKusick, Chapter 2, this volume.
5. D. E. Koshland, "Sequences and Consequences of the Human Genome," *Science* 246:189, 1989.
6. *Anna J. v. Mark C.*, 286 Cal. Rptr. 369 (Ct. App., 4th Dist., 1991).
7. R. N. Proctor, *Racial Hygiene: Medicine under the Nazis,* Harvard University Press, Cambridge, MA, 1988.
8. M. Foucault, *Histoire de la Sexualité: la Volonté de Savoir,* Vol. 1, Gallimard Ed., Paris, 1976, p. 183.
9. L.I.A. Bieke, "Is Homosexuality Hormonally Determined?," *Journal of Homosexuality,* 6:35–43, 1981; R. C. Lewontin, "Biological Determinism as a Social Weapon. In *Ann Arbor for the People*, Burgess Press, Minneapolis, 1977; E. O. Wilson, "Human Decency is Animal," *New York Times Magazine,* October 12, 1975, p. 50; C. Holden, "Study of Terrorism Emerging as International Endeavor," *Science* 203:33–5, 1979; "Math Genius May Have Hormonal Basis," *Science* 222:1312, 1983; and M. Konner, "The Aggressors," *New York Times Magazine,* August 14, 1988, p. 33.
10. A. R. Weinberg, "The Molecule of Life," *Scientific American* 3(1):92–101, 1991.
11. G. Lanteri-Laura, *Histoire de la Phrénologie,* Presses Universitaires de France, Paris, 1970. Even though this work focuses mainly on Franz Joseph Gall, the founder of phrenology (the mapping of the brain), the author presents an excellent analysis of anatomical reductionism before and after Gall.
12. R. Descartes, *Oeuvres de Descartes,* Vol. 11, Adam-Tannery, Paris 1910, p. 119.
13. R. Descartes, 1910, Vol. 6, pp. 61, 63.
14. *Ibid.*
15. E. Shuster, *Le Médecin de Soi-Même,* Presses Universitaires de France, Paris, 1972, Chapter 1, p. 15.
16. R. Descartes, *Lettre à Regius,* January 1642, Adam-Tannery, Paris, Vol. 3, pp. 563–5. See also, *Lettre à Mesland,* February 9, 1645, Adam-Tannery, Paris, Vol. 4, pp. 165–7, and *Les Passions de l'Ame,* art. 30, Adam-Tannery, Paris, Vol. 11, p. 351.
17. Michal Polanyi expressed similar views when he stated that "both machines and

living mechanisms are irreducible to the laws of physics and chemistry." A machine "as a whole, works under the control of two principles. The higher one is the principle of the machine's design, and this harnesses the lower one, which consists in the physical-chemical processes on which the machine relies." Similarly, an "organism is shown to be, like a machine, a system which works according to two different principles: its structure serves as a boundary condition harnessing the physical-chemical processes by which its organs perform their functions." Ultimately, for Polanyi, any structured system, specifically the structure of machines and the morphology of living things, "transcends the laws of physics and chemistry"; in D. Hull, *Philosophy of Biological Science,* Prentice-Hall, Englewood Cliffs, NJ, 1974, p. 139.

18. F. Jacob, *La Logique du Vivant. Une histoire de l'hérédité,* Gallimard ed., Paris, 1970, pp. 10, 45.

19. V. A. McKusick, "The Human Genome through the Eyes of a Clinical Geneticist," *Cytogenetics and Cell Genetics,* 32:7–23, 1982. See also V. McKusick, "Mapping and Sequencing the Human Genome," *New England Journal of Medicine* 320(14):910–5, 1989.

20. A. Vesalius, *De Fabrica Humani Corporis,* libri septum, Basel, Switzerland, 1543.

21. Aristotle, *De Partibus Animalium* 686a 24–25.

22. G. Canguilhem, *Etudes d'Histoire et de Philosophie des Sciences. L'Homme de Vesale dans le Monde de Copernics: 1543,* Librairie Philosophique J. Vrin, Paris, 1968, p. 31.

23. J. C. Stephens, M. L. Cavanaugh, M. I. Gradie, et al., "Mapping the Human Genome: Current Status," *Science* 250:237, 1990. This issue contains a special feature, "The Human Genome Map 1990" (pp. 262a–p). The cover illustration is a clever representation of the Vesalian man superimposed on genetic maps and charts to explore the new "Genetic Frontier." This illustrates the ideas I wish to convey.

24. H. L. Kaye, *The Social Meaning of Modern Biology,* Yale University Press, New Haven, CT, 1986, p. 55.

25. Jacques Monod, in H. F. Judson, *The Eighth Day of Creation,* Simon & Schuster, New York, 1979, p. 210.

26. Jacob, 1970, pp. 196–319.

27. S. Brenner, "New Directions in Molecular Biology," *Nature* 248:785–7, 1974.

28. W. C. Zimmerli, "Who Has the Right to Know the Genetic Constitution of a Particular Person?" In *Human Genetics Information: Science, Law and Ethics,* Ciba Foundation Symposium 149, John Wiley and Sons, Chichester, 1990, p. 93.

29. J. Rifkin, *Algeny: A New Word—A New World,* Penguin Books, New York, 1984, p. 208.

30. Jacob, 1970, pp. 10, 283.

31. Ibid., pp. 272–4.

32. Ibid., p. 208.

33. Kaye, 1986, p. 56.

34. Ibid.

35. B. McClintock, "The Significance of the Genome to Challenge," *Science* 226:792–801, 1984.

36. Kaye, 1986, p. 76.

37. G.J.V. Nossal, *Introduction to Human Genetic Information: Science, Law and*

Ethics, Ciba Foundation Symposium 149, John Wiley and Sons, Chichester, 1990.
38. F. Jacob, *Le Jeu des Possibles,* Fayard Ed., Paris, 1981, p. 126.
39. T. J. Bouchard, D. T. Lykken, et al. "Sources of Human Psychological Differences: The Minnesota Study of Twins Reared Apart," *Science* 250:223–8, 1990.
40. McClintock, 1984.
41. T. Dobzhansky, *Genetics of the Evolutionary Process,* Columbia University Press, New York, 1970, p. 36.
42. V. McKusick, 1982.
43. Jacob, 1970, p. 298.
44. L. Prigogine and I. Stengers, *Order out of Chaos: Man's New Dialogue with Nature,* Bantam Books, New York, 1984, p. 225.
45. R. Descartes, *Lèttre à Mersenne,* February 20, 1639, Adam-Tannery, Paris, Vol. 2, pp. 525–26.
46. Eric Lander, personal communication.
47. Renée Fox, personal communication.
48. Kaye, 1986, p. 72.
49. Jacob, 1970, p. 345.

7

If Gene Therapy Is the Cure,
What Is the Disease?

Arthur L. Caplan

It may strain credulity to say that the analysis of concepts such as health, disease, or normality can shed any light on the ethical and policy issues associated with the Human Genome Project. However, credulity must be strained. The understanding that our society and others have of the concepts of health, disease, and normality will play a key role in shaping the application of emerging knowledge about human genetics. If this is so, then why is it so difficult to direct the attention of those interested in the ethical, legal, and social consequences of new knowledge created by the Human Genome Project toward the analysis of these concepts.

Can Philosophical Analysis Really Help?

One reason is that many of those who deliver health care are not beset by doubts or ambiguities about the aims, goals, definitions, or purposes of their activities. Most people who seek health care do so because they believe that something is wrong with them. The aim of the doctor or nurse is clear—fix the problem by correcting anomalies, reversing pathological processes, or, if cure is not possible, providing some means of accommodation and palliation. In most interactions between health care provider and patient, both parties reach agreement that something is wrong and on what the goal of consequent medical intervention should be.

Even when the goal of medical intervention is the subject of disagreement between provider and patient, for example in making decisions about undertaking continued resuscitation efforts or kidney dialysis for imminently dying terminally ill patients, or the prolonged, intense utilization of life-sustaining treatments for extremely elderly patients who have been in permanent comas for many years, health care providers and patients, families, or guardians are almost always able to reach an agreement about what should be done.

Unless financing is an obstacle, it is almost always the case that when patients want treatment that has some chance of being effective doctors will agree to provide it. In American medicine, respecting autonomy is the core principle used to settle disputes about the ultimate goals that ought to guide the provision of care. In situations where doubt as to the goal of care exists, adherence to the principle of autonomy requires clinicians to defer to the wishes and preferences of those receiving care or their surrogates even when patient choice results in the forgoing of likely benefits.[1] The rule "yield in the face of autonomy" minimizes the need to carefully examine more general questions about the aims and goals of health care.

Still another reason for the relative absence of discussions of the definition of health and disease is that when there is a lack of consensus about the appropriate classification of particular traits, states, or behaviors as constituting instances of either disease or health, the disagreements sometimes become so heated that there is little incentive to join what may appear to more closely resemble a fray rather than a discussion. There is still disagreement about the scope or application of the concepts of health, disease, or normality to gambling, sexual promiscuity, premenstrual syndrome, hyperactivity, or homosexuality. The battles over the classification of these behaviors and traits have been heated and fierce.[2] Uncertainty about whether or not to attempt to treat short stature in children, low blood sugar, and hypertension has also led to heated controversies as to their disease or health classification and, consequently, the appropriateness or inappropriateness of therapeutic intervention. Controversy about the scope and domain of the concepts of health and disease sometimes is so divisive that it may seem prudent to some to simply avoid asking questions about the application of concepts such as health and disease to new knowledge in the area of human heredity.

There is one other reason why the implications of new knowledge in genetics for understanding health, disease, and normality are not always centerstage. Those who are actually engaged in mapping and sequencing the human genome, or the genomes of other organisms, often do not have any particular practical goal or application motivating their work. Despite all the hand wringing that has accompanied the evolution of the Genome Project, and the promises of therapeutic benefits that it will produce, many of those involved are simply interested in understanding the composition of the genome, its infrastructure or anatomy. Basic researchers have fewer reasons than clinical researchers to struggle to clarify the conceptual foundations of health, disease, and normalcy, and they will, of necessity, control the direction of the Human Genome Project for many years to come.

Not only do basic researchers have less reason to explore the conceptual foundations of health and disease, in the case of the Genome Project they have a positive disincentive to participate or encourage such an inquiry. If uncertainty about what to do with new knowledge in the realm of genetics is a cause for concern in some quarters, then those who want to proceed quickly with mapping the genome might find it prudent to simply deny that any application of new knowledge in genetics is imminent or to promise to abstain from any controversial applications of this knowledge.

Promising to try and avoid doing anything that will grossly offend societal mores is the simplest strategy if one's aim is not applying new knowledge but merely to be allowed to proceed to acquire it. Some involved in the Genome Project have tried to defuse worries about the thorny issues of what constitutes health, what are the boundaries of normality, and what justifies referring to a particular genetic state as a disease by self-imposed restrictions on the application of new knowledge concerning the genome. The clearest example of this prophylactic strategy is the promise that germline gene therapy will not be done.[3] The messy problem of how to fit new knowledge about heredity into existing categories of disease, normality, and health can, perhaps, be forestalled by arguing that the sole therapeutic goal of the Human Genome Project is somatic genetic therapy for obvious, clear-cut instances of human disease.[4]

What Is Disease?

The easiest way to start an examination of how new knowledge of the human genome will influence how we should understand the goals of health care is by asking what is it today that doctors, nurses, psychologists, physical therapists, and the myriad other professionals who work in health care are supposed to do? The most obvious and common-sense answer is that they are to combat disease. Although there are many other goals that might and have been added, ranging from screening for eligibility for government benefits to certifying persons as fit to play sports or serve in the military, the fight against disease occupies center-stage in what people expect health care providers to do. So if it is possible to become clearer about what disease is, then it may be possible to have a better understanding of the boundaries of what is and is not licit with respect to the application of new knowledge arising from the Human Genome Project.

There are two major points of contention evident in the literature on the meaning of the concept of disease.[5] One major source of disagreement concerns the role played by the determination of normality in the identification of disease. The other concerns the role played by values in the definition of disease.

Many physicians and nurses equate difference with disease or, at least with a reason for suspicion that disease may exist. To put the point another way, abnormality is often viewed with suspicion either because it is seen as disease or because it is seen as symptomatic of an underlying state of disease. As E. A. Murphy cogently observed, "the clinician has tended to regard the disease as that state in which the limits of the normal have been transgressed."[6]

For example, many physicians and public health experts believe that blood pressure readings that vary from what is considered typical or normal for specific age groups within the population are, in themselves, indicative of disease. For American physicians, variations skewed toward high numbers are disturbing. For German physicians, both high and low numbers are equally likely to be diagnosed as disease and, therefore, as sufficient to merit medical intervention. Similarly, variations, even of a modest sort, with respect to height, weight, attention span, sperm production, or blood cholesterol levels may trigger disease labels and consequent efforts at therapy.

Critics of what can be called the "disease as abnormality" approach point out that there is nothing inherent about difference that makes a particular biological, chemical, or mental state a disease. Moreover, since variation is an omnipresent feature of human beings, it is especially odd to argue that extremes of variation are somehow indicative of disease. If there is nothing at all unnatural about variation, then abnormality cannot in itself be equated with disease. Indeed, critics of the view that equates difference with disease note that this equation has throughout the history of medicine led to the classification of differences with respect to race, gender, and ethnicity as diseases, which in turn has been the basis for unfair and even harmful interventions against persons suffering from nothing more than a darker skin color or the presence of ovaries.[7]

Those who are skeptical of the equation of disease with abnormality make two points worth pondering in the context of new knowledge about human heredity. Simply labeling abnormality or difference as disease is, of necessity, to impose a value judgment on a physical or mental state that does not wear its disease status on its sleeve. Second, abnormality is not in itself bad. After all, those who are unusually smart, strong, fast, or prolific are not classified as diseased. If abnormality is to be equated with disease, then at a minimum it must be abnormality that is associated with something that is disvalued and, further, there must be some connection between the difference or abnormality and the dysfunction that is disvalued.[8]

These criticisms of the equation of abnormality and disease raise another major point of contention in defining the concepts of health and disease—whether it is possible to do so without reference to values, for if values must be used to decide whether a particular abnormality or difference is indicative of disease, then many worry that the entire process of defining health and disease must be subjective and especially vulnerable to political or social influences.[9] If disease refers to abnormal states, either mental or physical, that are disvalued, then the appearance of values seems to some to make the prospects grim for objectivity or consensus about what states are or are not healthy or diseased. Subjectivity and a lack of consensus could bode especially ill for the uses to which new knowledge of human heredity might be put since applications might be controlled by the powerful or the economically privileged to advance their own values. Some believe that values need not be invoked in defining health or disease. Those who espouse nonnormativism, the view that the definition of disease need not involve any invocation of values, usually invoke some notion of natural function or design to ground the definition of disease.[10]

Nonnormativists doubt that values must enter into the assessment of organ function and behavior.[11] For example, if a cardiologist says that the function of the heart is to pump blood or a renal physiologist claims that the function of the kidney is to cleanse the blood of impurities, it is not because they hold certain values about hearts or kidneys. Rather, based on both functional and evolutionary analysis, it is possible to arrive at an understanding of what the organs are supposed to do. By removing both kidneys or seeing what happens if the heart is damaged, it is possible to ascertain the functions of these organs. Therefore, nonnormativists argue, it is possible to utilize concepts such as cardiac or renal disease without invoking any sort of value judgments as to whether or not it is good that hearts beat or commendable that kidneys filter the blood.

Normativists argue that the concept of disease is inherently and inextricably value-laden.[12] They believe that functional analysis by itself cannot reveal whether a particular state of the body or mind is indicative of disease. For example, the fact that someone is nearsighted or farsighted may or may not be indicative of disease or disability. It depends on whether one is going to spend one's day in the library, in the operating room, or hunting on the savannah.

Even the failure of a major organ may not be an obvious instance of a disease if other options exist for performing the same function. The fact that a kidney can no longer cleanse the blood of impurities may or may not be indicative of disease, depending on such value-laden judgments as whether its owner wants it to do so. Artificial kidneys can supplant the function of natural ones, so there is nothing inherent about the desirability of having a natural kidney cleanse the blood. It is possible that some people might prefer to have their blood cleansed by a machine (maybe a more efficient and safe futuristic one than is now currently available) to having their own kidneys do the job. Consequently, normativists say, the view that a kidney incapable of cleansing the blood of impurities is diseased is as much a claim about values and options as it is about renal physiology.

Normativists almost always subscribe to the view that assessments of health and disease are value-laden and that as a result they are inherently subjective, not objective. However, the link between the presence of values and the threat of subjectivity is open to question. The presence of values may make the definition of disease or health suspect in terms of its objectivity. But the existence of values in the assessment of health and disease does not mean that it is impossible to reach consensus about the definition of disease despite the fact that values play a role in the definition.[13]

Nonnormativism versus Normativism: Finding a Middle Ground

Equating normality with health and abnormality with disease as ways of defining these key concepts seems open to devastating conceptual and empirical problems. The view that what is different is, in itself, disease simply does not square with ordinary experience. Many differences are viewed as desirable or beneficial. Difference may be a cause for inquiry into whether or not a particular state is indicative of health or disease, but it is not in itself sufficient as a basis for deciding whether abnormality or variation represents health or disease in an individual or group. This is especially true for genetic differences, where it is already known that wide variations exist in terms of the genetic makeup of individuals that are not in any way manifest in their overt features, traits, or behaviors. Difference in itself does not always make a difference. However, the definition of disease and health does seem closely tied to those differences or abnormalities that are disvalued by the individual or group. If a particular trait or behavior or physical structure is seen as causing impairment, dysfunction, pain, or other disvalued states, then it is a prime candidate for categorization as a disease.

The problem with linking values and abnormalities is that not all states commonly recognized as diseases are necessarily indicative of difference or abnormality. Nor are all dysfunctions or impairments always disvalued. For example, every human being may suffer from the common cold, acne, anxiety, or dental caries. The universality of these conditions does not make them any more palatable or any the less disvalued. It is possible for a state to be typical, normal, common, or universal and nonetheless be sufficiently disvalued that it is classified by both lay persons and health care professionals as disease.

Not every dysfunctional or impaired state is disvalued. Those who do not wish to have children may rejoice to discover that they have ovaries that are incapable of ovulation, lack a uterus, or have testes that cannot create sperm. Someone born with only one functioning kidney may remain entirely indifferent to and even unaware of this dysfunction. Not all dysfunctional states are necessarily disvalued, meaning that not every abnormal state can be viewed as a disease. Nevertheless, there does appear to be a conceptual link between abnormality, dysfunction, and disvaluation. If we restrict the definition of disease to cover only those mental or physical states of human beings that are abnormal, dysfunctional, and disvalued, then we will be able to identify many of the diseases recognized within a group or in a society.

Two questions confront those who admit the tie between disvaluation and dysfunction. Is it possible to determine the existence of dysfunction without invoking value judgments? If not, does the presence of disvaluation as a key criterion of the definition of health and disease make these concepts so subjective as to be either useless or extremely vulnerable to abuse by those powerful enough or privileged enough to impose their personal values on others?[14]

XYY and Oculocutaneous Albinism: Are These Diseases?

There is no need to try to resolve the disputes about normality or the role of values in the definition of disease in order to see what the consequences of these disputes are for new knowledge that is arising and will, it is hoped, continue to arise in the domain of human genetics. The stance that those in clinical genetics adopt toward assessing the significance of difference and abnormality at the level of the genome, the role of values in defining genetic disease, and the need to link genetic disease to dysfunction will play pivotal roles in what is done with the knowledge generated by ongoing work to map and sequence the genome.

How disease is currently assessed in the realm of clinical genetics is not entirely a hypothetical question. After all, counselors and clinicians have been treating patients for genetic diseases for decades. It is instructive to look and see how they currently define disease and health in order to try and forecast how new knowledge about human heredity will be absorbed into clinical and counseling practices. Two forms of genetic abnormality illustrate how much uncertainty and confusion exists about both the criteria that ought be used to define disease and the proper application of the concepts of health and disease at the level of genetic difference and abnormality.

Not long ago, a woman at a large medical center was informed by a genetics counselor that the fetus she was carrying possessed an abnormal chromosome. The child had XYY syndrome—an extra Y chromosome. The mother and father had sought genetic testing because she was in an age group at somewhat higher risk for a different genetic condition, Down syndrome. The information that the baby had a different chromosomal abnormality from the one that had concerned the parents came as a bit of a surprise. Subsequently, in counseling sessions, the mother and father were told that a few researchers had posited a connection between criminality and this chromosome abnormality. They were also told that some researchers believed there was also a link between this condition and tall physical stature and even severe acne. After talking about the situation with their family doctor and various friends, the couple decided to abort the pregnancy. Is XYY syndrome a disease? If not, why were the parents told it had been detected? And if it is, is it a disease that merits aborting a fetus with this condition?

Oculocutaneous albinism (OCA) is a disorder in which melanin is absent or decreased in the skin, hair, and eyes. Albinism actually refers to a group of autosomal recessive traits in which the enzyme necessary for melanin production, tyrosinase, is present or absent in varying degrees, causing a fair degree of heterogeneity. Most forms of OCA are associated with a very distinctive set of complications.[15] Tyrosine-negative albinism, which occurs at a rate of about one in every 34,000 births, is associated with extreme sensitivity to light (photophobia), nystagmus, severe impairment in visual acuity, and a greatly increased risk of squamous cell skin cancer.

The genes involved in OCA are rapidly being mapped.[16] Prenatal diagnosis is already a possibility, using fetoscopy to obtain samples of fetal skin or scalp hair around the sixteenth week of development to see whether the follicles contain melanin, but the relatively low incidence of the condition and the costs involved mean that such testing is rarely done. It will not be long, however, before a routine test will be available to detect this genetic abnormality from any sample of fetal DNA. But, unless albinism is a disease, why should anyone try to detect it, much less provide information about it to parents?

Decisions about what makes a condition a disease have direct implications for what will be done with information about XYY syndrome or OCA. If one subscribes to the definition of disease as abnormality, then both conditions will constitute disease states and both ought be reported as such to parents. If one believes that it is necessary to draw a connection between disease and dysfunction, then OCA will certainly qualify as a disease whereas, in lieu of more evidence, XYY syndrome may not. And if one takes the position that to be a disease a state must be both dysfunctional and disvalued, then it is possible that neither XYY syndrome nor OCA qualifies as a disease.

My own view is that it is difficult to defend the decision to label either XYY syndrome or OCA as a disease. The former is an abnormality for which there is no reliable evidence that it causes or is associated with dysfunction. The latter is an abnormality that produces dysfunction, but the problems associated with the condition are readily amenable to various interventions and coping strategies. It seems accurate to say that a person with OCA is not diseased, but has an abnormality that causes some impairment.

It should be obvious that the decisions that are made regarding the application of the concepts of health and disease in the realm of genetics require explanation and justification. It should also be obvious that not everyone will arrive at the same conclusion about the classification of human genetic differences because different views may be held about what justifies classifying a trait or characteristic as a disease, an abnormality, or a healthy state. Serious consequences follow the determination of disease states. Those in clinical genetics who diagnose and treat genetic diseases must not invoke the oft-espoused desire to remain value-neutral as a rationale or an excuse for avoiding the obligation to carefully think through the criteria that are now used, or others that might be used in the future, to classify genetic diversity and differences as indicative of health or disease.

Implications of Defining Disease for the Study of the Human Genome

The realization that there is a broad spectrum of opinion about what makes something a disease is troubling in speculating about the ways in which new knowledge about the human genome is likely to be analyzed and classified. Much of the information likely to be acquired will reveal far more about the structure and composition of the genome than about its function. If we forget this, there is a grave danger that the disease-as-abnormality approach will find fertile ground in the realm of genetics in the future.

Some agreement must be reached about whether it is necessary to establish a link between genetic information and dysfunction, or between dysfunction and disvaluation, in order to establish the disease status of particular bits of the genome. Otherwise, increased screening and testing will reveal more and more differences and variations among our genomes, which will lead to an incoherent set of responses in terms of counseling, reproductive choices, and therapy. Although consistency need not always be desirable, it seems morally incumbent on those who will be faced with the challenge of applying new knowledge about the genome to strive for some sort of professional and societal consensus as to how these questions ought be answered.

The first wave of new information about the human genome is likely to include a fair amount of news about human differences that may have uncertain or unknown phenotypic consequences. Clinical genetics is still in its infancy. As such, it must proceed with great caution in labeling states or variations as abnormal, much less diseases. For now, clinical genetics should restrict itself to the identification and assessment of only those genetic states which are known to be dysfunctional as well as different. It should discourage efforts to allow "fishing expeditions" to become part of prenatal, carrier, or workplace screening. And, it should assert clearly that the central goal of human clinical genetics is the prevention or amelioration of disease, not the improvement of the genome. It is important to note that abjuring eugenics as a proper goal of clinical genetics is not the same thing as forgoing any effort to meddle or intervene with the genetics of reproductive cells.

Please Leave Us Alone, We Promise to Be Good!

Those who believe in the value of the Human Genome Project, a group to which I belong, often try to calm fears about the misapplication of the knowledge the Project intends to create by assuring everyone that their intentions are pure. This promise meets the concern of some of the harshest critics of the project.[17] A few of those who doubt that humankind knows what to do with more information about its own hereditary makeup or who simply believe that it is unnatural to mess around with genes sometimes try to arouse legislative or public concern by spinning scenarios in which human-animal chimeras slink out of the corridors of MIT, Cal Tech, Genentech, or Fort Dietrich to commit maniacal human-animal misdeeds against hapless humans. If such grim scenarios aren't scary enough, the occasional critic resorts to even more horrifying futuristic timeworms in which hordes of clones derived from the embryos of businessmen, sports stars, and politicians (no attempt is made to mitigate the horror) descend on an unsuspecting and defenseless world. In the most hyperventilating form of such criticism, warnings are issued that if the Genome Project is not stopped now the result will inevitably be a planet teeming with millions of knockoff copies of Adolph Hitler, Genghis Khan, Saddam Hussein, Idi Amin, and Joseph Stalin.

Those who want the Genome Project to proceed are quite willing to promise never to xenograft a gene or clone Adolph Hitler. If, by promising not to clone or even to perform germline interventions, the defenders of the Genome Project can allay the fears of their most strident and publicly visible critics, they are more than happy to do so.[18] This is especially true when the science fiction scenarios being spun are either scientifically impossible (cloning Hitler) or well beyond the reach of contemporary science (gene xenografting).

The greatest challenge to securing continuing funding for the Genome Project does not originate from concerns about privacy, confidentiality, or coercive genetic testing. It is eugenics, manipulating the human genome to improve or enhance the human species, that is the real source of worry. This is reflected in the content of the futuristic horror scenarios spun by the Project's critics.[19] It is also rooted in the historical reality of social policies based on eugenics that led to the deaths of millions in this century.[20]

Promising not to do anything that remotely hints of germline engineering, where eugenics is the goal, is relatively easy for those connected with the Genome Project because none of them believe that anyone is even remotely close to knowing how to alter the germlines of a human being, much less whether germline engineering will actually work. If the situation were different with respect to the practicality of germline interventions, then statements to the effect that germline intervention ought never be attempted might be far more muted. Even without the prospect of imminent application, they should be. The promise never to do germline engineering, invoked merely as an expedient way to silence critics, is implausible because it rests on a flimsy moral foundation. Why shouldn't a couple concerned about passing along hemophilia or sickle cell disease hope that medicine can help alter their genomes so as to minimize their risks of doing so? Why shouldn't clinicians fervently want to undertake some forms of germline interventions so as to eliminate diseases such as Tay-Sachs, thalassemia, or Hurler syndrome?

If it were possible to eliminate a lethal gene from the human population by germline alterations, is there any convincing moral reason why this should not be done? If those carrying a lethal gene request treatment so they are able to reproduce without guilt or fear, ought not health care providers feel not reluctance, but a duty to help them? If the prevention and treatment of disease are the goals of human clinical genetics, then not only should germline therapy not be forgone, it may be morally obligatory in cases where no somatic therapy is possible.

Should Germline Interventions Be Forgone?

Some believe that any attempt at germline therapy is wrong because it requires imposing the risk of harm on future generations either by causing unanticipated side effects in unborn infants or by introducing dangerous genes into the gene pool. Future persons have no say in whether or not they consent to having risks imposed on them. They, not their parents or ancestors, will suffer should attempts to manipulate the germline produce untoward results. Many bioethicists believe, and existing government policy in the United States, Germany, and other nations maintains, that it is wrong to impose the risk of serious harm on those who cannot themselves consent.[21] Newborns, very young children, the severely mentally ill, and the severely mentally retarded should simply not be recruited as the subjects of research. The other major reason for not undertaking germline interventions is that hereditary information that is of value, not for the individual but for the species, may be lost. If lethal or disabling genes are removed from a certain individual's gametes, it may be that benefits conferred on the population when these genes recombine with other, nonlethal genes will be lost.

Another argument against germline therapy is that no one would really want to use it for the purposes of eugenics. But this is patently false. Even putting aside Germany's three-decades-long embrace of race hygiene and eugenics (see Chapter 4, this volume), there are examples in our own time of governments and private organizations avidly and unashamedly pursuing eugenic goals. The government of Singapore instituted numerous eugenic policies during the 1980s, including a policy of providing financial incentives to "smart" people to have more babies.[22] The California-based Repository for Germinal Choice, known more colloquially as the Nobel Prize sperm bank, has assigned itself the mission of seeking out and storing gametes from men selected for their scientific, athletic, or entrepreneurial acumen. Their sperm is made available for use by women of high intelligence for the express purpose of creating genetically superior children who can improve the long-term happiness and stability of human society. Few protests have greeted these activities whereas the hypothetical suggestion of someday directly modifying the genetic blueprint of a sperm or an egg has elicited great concern in many quarters. Granted, eugenics has been horribly abused in the past and may still result in terrible abuses today. But, it is simply a confusion to equate eugenics with any discussion of germline therapy.

Should scientists or clinicians really promise never to try to eliminate or modify the genetic messages contained in a sperm or an egg if that message contains instructions that may cause sickle cell disease, Lesch-Nyhan syndrome, or retinoblastoma? The grim history of eugenically inspired social policy tells why it is important to protest and even prohibit the activity of the Nobel Prize sperm bank or to vehemently criticize the birth incentive policies of Singapore.[23] It does not provide an argument against allowing voluntary, therapeutic efforts using germline manipulations to prevent certain and grievous harm from befalling future persons. There is no slope that leads inexorably from therapeutic germline interventions intended to benefit future persons to the creation of eugenically driven, genocidal social policies. Nazi eugenic policies were not aimed at benefitting individuals. The state or the *Volk,* not the individual, was the object of Nazi eugenic policy. Public health, not individual therapy, was the driving force behind the Nazi medicalization of eugenics.

Worries about imposing harms on future persons without their consent or robbing the gene pool of the value of diversity are even less persuasive reasons for foreswearing germline intervention. If the harm that will befall the as yet unborn person are serious, even fatal, then it is far from self-evident that it would be wrong to try to prevent the harm even if it means the imposition of possible risk on the child who will be born. The risk must be such that the child who is the involuntary subject of germline experimentation is at grave danger of being made worse off by the intervention than would have been true had the child been born with no effort to alter his or her genetic defect. Some genetic diseases are so miserable and awful that at least some genetic interventions with the germline seem justifiable.

It is at best cruel to argue that some people must bear the burden of genetic disease in order to allow benefits to accrue to the group or species. At best, genetic diversity is an argument for creating a gamete bank to preserve diversity. It is hard to see why an unborn child has any obligation to preserve the genetic diversity of the species at the price of grave harm or certain death. Forgoing efforts at germline engineering makes some political sense in the current climate of concern about the Genome Project. But, it makes no sense conceptually or ethically. The danger inherent in such stances is that they will result in important benefits being delayed or lost for persons who have impairments or diseases that might be amenable to germline engineering.

The way to handle legitimate concerns about the dangers and potential for abuse of new knowledge generated by the genome is to forthrightly examine what are and are not appropriate goals for those who provide services and interventions in health care. There is nothing sacrosanct about the human genome. It is only our inability to openly and clearly define what constitutes disease in the domain of genetics that makes us feel that intervention with the germline is playing with moral fire. If it is eugenics we abhor, then it is eugenic goals which should be forgone.

NOTES

1. J. Childress, *Who Should Decide?*, Oxford University Press, New York, 1982.
2. For example, A. Caplan and T. H. Engelhardt, eds., *Concepts of Health and Disease,* Addison-Wesley, Reading, MA, 1981.

3. W. F. Anderson, "Human Gene Therapy: Why Draw a Line?" *Journal of Medicine and Philosophy* 14:681–93, 1989.

4. Ibid. and Annas and Elias, Chapter 1, this volume.

5. Caplan and Engelhardt, 1981, and A. Caplan, "The Concepts of Health and Disease," in *Medical Ethics,* edited by R. Veatch, Jones and Bartlett, Boston, 1989, pp. 49–63.

6. E. A. Murphy, *The Logic of Medicine,* Johns Hopkins University Press, Baltimore, 1976, p. 122.

7. V. N. Gamble, "Race, Gender and Class in American Medicine," speech at a symposium on Race, Prejudice and Health Care: The Legacy of the Tuskegee Study, Minneapolis, MN, June 1, 1991.

8. K. Clouser, C. Culver, and B. Gert, "Malady: A New Treatment of Disease," *Hastings Center Report* 11(3):29–37, 1981.

9. Caplan, 1989.

10. For example, C. Boorse, "On the Distinction between Disease and Illness," *Philosophy and Public Affairs* 5:49–68, 1975; C. Boorse, "What a Theory of Mental Health Should Be," *Philosophy of Science* 44:542–73, 1976; C. Boorse, "Concepts of Health," in *Health Care Ethics,* edited by D. Van De Veer and T. Regan, Temple University Press, Philadelphia, 1987, pp. 359–94.

11. For example, Clouser, Culver, and Gert, 1981.

12. For example, H. Fabrega, "Concepts of Disease: Logical Features and Social Implications," *Perspectives in Biology and Medicine.* 5:538–617, 1972; P. Sedgwick, "Illness—Mental and Otherwise," *Hastings Center Studies* 1:19–40, 1973.

13. For example, Clouser, Culver, and Gert, 1981, and A. Flew, "Mental Health, Mental Disease, Mental Illness: The Medical model," in *Mental Illness,* edited by P. Bean, John Wiley and Sons, New York, 1983.

14. See Caplan and Engelhardt, 1981.

15. R. Abadi and E. Pascal, "The Recognition and Management of Albinism," *Ophthalmology and Physiological Optometry* 9:3–15, 1989.

16. R. Spritz, K. Strunk, L. Giebel, and R. King, "Detection of Mutation in the Tyrosinase Gene in a Patient with Type IA Oculocutaneous Albinism," *New England Journal of Medicine* 322:1724, 1990.

17. J. Rifkin, *Declaration of a Heretic,* Routledge & Kegan Paul, Boston, 1985; J. Rifkin, "Perils of Genetic Engineering," *Resurgence* 109:4–7, March/April 1985.

18. For example, S. S. Hall, "James Watson and the Search for Biology's 'Holy Grail'," *Smithsonian,* February 1990, pp. 41–51.

19. J. Swazey, Chapter 3, this volume.

20. R. Proctor, Chapter 4, this volume.

21. For example, L. Kass, "Regarding the Ends of Medicine and the Pursuit of Health," *The Public Interest* 40:11, 1975; P. Ramsey, *The Patient as Person,* Yale University Press, New Haven, CT, 1970.

22. C. K. Chan, "Eugenics on the Rise: A Report from Singapore," *Ethics, Reproduction and Genetic Control,* edited by R. Chadwick, Routledge, London, 1987, pp. 164–72.

23. T. Duster, *Backdoor to Eugenics,* Routledge, New York, 1990.

Somatic and Germline Gene Therapy

Sherman Elias and George J. Annas

Recent reports of human gene therapy experiments have been heralded worldwide with praise from the medical and scientific communities as well as the general public. A new medical journal, *Human Gene Therapy,* devoted entirely to this topic has also been introduced. The stage is now set for more experiments involving human somatic cell genes with the objective of ameliorating or correcting genetic defects.[1] In mid-1990 Steven Rosenberg and colleagues reported the use of retroviral-mediated gene transduction to introduce the gene coding for resistance to neomycin into human tumor-infiltrating lymphocytes before their infusion into five patients with metastatic melanoma.[2] The distribution and survival of these "gene-marked" lymphocytes were then studied, and for the first time, the feasibility and safety of introducing new genes into humans were demonstrated. We are now witnessing the transition from theoretical possibility to practical reality of the "dawn of a new age of cancer treatment" based on novel gene therapy strageties.[3] In September 1990, Michael Blaese and colleagues at the National Institutes of Health (NIH) announced the first human gene experiment in a young girl with adenosine deaminase (ADA) deficiency, a very rare autosomal recessive genetic disorder resulting in severe combined immune deficiency.[4] Affected individuals usually die during childhood from chronic infections. The investigators obtained circulating lymphocytes from the patient, expanded the number of cells in culture, infected the cells with a recombinant retrovirus carrying the ADA gene, and then reinfused the cells into the patient.

Somatic Cell Therapy

Some believe that somatic cell gene therapy is an unwise pursuit. Their underlying concern is that it will inevitably lead to the insertion of genes to change the character of people, reduce the human species to a technologically designed product, or even "change the meaning of being human."[5] These fears rest on the presupposition that somatic cell gene therapy differs in a significant way from accepted forms of treatment.[6] This has not, however, been the consensus of panels of experts who have analyzed the issues. Both the President's Commission for the Study of Ethical Problems in Medicine and Biomedical and Behavioral Research[7] and the European Medical Research Councils[8] concluded that somatic cell gene therapy is not fundamentally different from other therapeutic procedures, such as organ transplantation or blood transfusion. The only important differences between somatic cell gene therapy and other treatments are the issues of safety and the possibility that viral vectors may also infect germ cells.

At present, among the various methods for introducing genetic material into cells, retroviral-based vectors appear to be the most promising approach for gene transfer in humans.[9] Concern has been raised that despite elaborate "built-in safety features" with redesigned mouse retroviral vectors, there is a finite risk that these vectors could recombine with undetected viruses or endogenous DNA sequences in the cell and so become infectious. This risk of "viral escape," as well as other potential risks that would be limited to the patient, such as activation of a proto-oncogene or disruption of an essential functioning gene, has resulted in society demanding very critical review and participation in any decisions to embark on human gene therapy.[10]

In the United States, recombinant DNA research funded by the NIH must first be reviewed by the local Institutional Review Board and local Institutional Biosafety Committee. In addition, the NIH has established the Recombinant DNA Advisory Committee (RAC) to review proposals to be funded by the NIH. An interdisciplinary subcommittee of RAC, called the Working Group on Human Gene Therapy, was formed that consisted of three laboratory scientists, three clinicians, three ethicists, three attorneys, two public-policy specialists, and one lay member. Based on a draft from this subcommittee, the RAC composed a document called "Points to Consider in the Design and Submission of Human Somatic-Cell Gene Therapy Protocols."[11] This document is intended to provide guidance in preparing proposals for NIH consideration and focuses on: (1) objectives and rationale of the research, (2) research design, anticipated risks and benefits, (3) selection of subjects, (4) informed consent, and (5) privacy and confidentiality. The RAC has deliberately limited its purview to somatic cell gene therapy, emphasizing that the "recombinant DNA is expected to be confined to the human subject" to be treated.[12]

W. French Anderson, a leading NIH investigator in the field of gene therapy, contends that "somatic cell gene therapy for the treatment of severe disease is considered ethical because it can be supported by the fundamental moral principle of beneficence: It would relieve human suffering."[13] He also argues that after the appropriate approval of therapeutic protocols by Institutional Review Boards and the NIH,

> it would be unethical to delay human trials. . . . Patients with serious genetic disease have little other hope at present for alleviation of their medical problems. Arguments that genetic engineering might someday be misused do not justify the needless perpetuation of human suffering that would result from unethical delay in the clinical application of this potentially powerful therapeutic procedure.[14]

And more recently, Anderson has insisted that it is time to be less tentative about somatic cell therapy, saying "What's the rush? The rush is the daily necessity to help sick people. Their (our) illnesses will not wait for a more convenient time."[15] A large majority of Americans seem to agree. According to a survey reported by the Office of Technology Assessment, 83 percent of the public say they approve of human cell manipulation to cure usually fatal genetic diseases.[16]

It should be emphasized, however, that human experimentation regulation has a long history, and it is easy to forget even its basic lessons in the rush to research. We are unlikely to repeat Martin Cline's unapproved beta-thalassemia experiments, but we do continue to concentrate early experiments on children (who cannot consent for themselves) and the terminally ill (who are especially vulnerable to coercion), repeating many of the same ethical mistakes made with artificial heart and xenograft transplants in the 1980s. The point is that just because somatic cell experiments are not fundamentally different than transplant experiments, it does not mean that there are no ethical or legal problems of consequence involving these experiments, or that we have reached a consensus on how to resolve them.

Germline Cell Therapy

In contrast to somatic cell gene therapy, which is limited to the individual patient, germline gene therapy entails insertion of a gene into the reproductive cells of the patient in such a way that the disorder in his or her offspring would also be corrected. This can be interpreted either as the corrective genetic modification of gametes (sperm or ova) or their precursor cells, or insertion of genetic material into the totipotential cells of a human conceptus in refined variations of the techniques used for germline modification in the creation of transgenic animals.[17] As such, germline gene therapy constitutes a definitive qualitative departure from any previous medical interventions because the changes would not only affect the individual, but also would be deliberately passed on to future generations.[18]

Should human germline gene therapy be permitted? Until now there has been some comfort in the fact that technical difficulties would preclude consideration of such genetic engineering in humans for many years to come. However, one need only look at the scientific advances presented at the First International Symposium on Preimplantation Genetics, in Chicago in September 1990, to realize that the pace of technology in this area has become much faster than previously predicted. Although in the past it may have seemed politically prudent to avoid the subject, we must now begin to seriously discuss the ethical issues before germline gene experiments in humans are technically possible in order to assist future policy makers in their deliberations.[19]

At a Workshop on International Cooperation for the Human Genome Project, in Valencia in October 1988, French researcher Jean Dausset suggested that the Genome Project posed such great potential hazards that it could open the door to Nazi-like atrocities. In an attempt to avoid such consequences, he suggested that the conferees agree on a moratorium on genetic manipulation of germline cells and a ban on gene transfer experiments on early embryos. The proposal won widespread agreement among the participants at the meeting and was only defeated after a watered-down resolution using "international cooperation" was suggested by Norton Zinder, who successfully argued that the group had no authority to enforce such a resolution. Leaving the question of authority for later debate, should we take Dausset's proposal for a moratorium on germline cell experimentation seriously?

Arguments against Human Germline Gene Therapy

The ethical arguments against the use of a human germline gene therapy fall into three categories: (1) its potential clinical risks, (2) the broader concern of changing the gene pool, the genetic inheritance of the human population, and (3) social dangers.[20]

Potential Clinical Risks. Major advances in knowledge must be made to overcome the significant technical obstacles for human germline gene therapy. Methods would have to be developed to stably integrate the new DNA precisely into the right chromosomal site in the appropriate tissues for adequate expression and proper regulation. The possibilities for dangerous mistakes are formidable: important examples would include insertional mutagenesis, in which a normally functioning gene is disrupted or a proto-oncogene activated by the newly integrated gene, or the regulatory signals of the nonfunctional or dysfunctional gene could adversely affect the regulation of the new exogenous gene.[21] These safety concerns are compounded by practical problems, including loss of gametes or embryos by instrument manipulation and the inherent limitations and inefficiency of reproductive technologies such as in vitro fertilization (IVF). Of course, many medical experiments have significant risks, and in this respect, germline research is not unique. IVF itself was opposed on similar grounds in the late 1960s and early 1970s.

On the other hand, unless there is overwhelming evidence that the procedure will be successful and not cause harm to the resulting child, there is no justification for doing genetic experiments on an early embryo and then reimplanting it. The more reasonable position would be not to reimplant the genetically defective embryo in the first place. The same logic applies to gamete manipulation. This argument suggests that except in the vanishingly rare case in which both parents are homozygous, and thus all embryos produced will be affected with their condition, and this condition is seen as unacceptable by them, *and* they refuse to use either donor ovum or donor sperm, there will be no clinical role for genetic therapy to replace or substitute for defective genes in embryos. Even in this case, of course, such research may never be justifiable because of the significant risks of introducing even worse problems for the resulting child.

Changing the Gene Pool. In germline gene therapy the objective would be to correct a genetic defect for future generations and hence restore the individual's lineage to the "normal" state. But what may be considered a harmful genetic trait today could be neutral in the future or could even conceivably serve a beneficial function, depending on environmental pressures in which the trait operates. Arno Motulsky has used the example of wearing eyeglasses. For primitive humans, genetically controlled myopia could prove fatal if one could not keep a sharp eye out for predators. The relatively high frequency of myopia and the need to wear eyeglasses represents loss of an adaptive biological trait in modern civilization. Nonetheless, myopia is never fatal, and some people, particularly ophthalmologists, optometrists, and eyeglass makers, might even say that the existence of myopia and other refractory deficits has been of benefit by creating a livelihood for them.[22] One danger is that we may wrongly eliminate a characteristic, such as sickle cell trait, that by protecting against malaria has advantages for the carrier. On the other hand, much of medicine has the potential for altering the next generation by helping the current one survive to reproductive age. As Alexander Capron has noted:

> The major reasons for drawing a line between somatic-cell and germ-line interventions . . . are that germ-line changes not only run the risk of perpetuating any errors made into future generations of nonconsenting "subjects" but also go beyond ordinary medicine and interfere with human evolution. Again, it must be admitted that all of medicine obstructs evolution. But that is inadvertent, whereas with human germ-line genetic engineering, the interference is intentional.[23]

Capron continues the argument by noting that "the results produced by evolution at any points in time are hardly sacrosanct," producing as they do genetic diseases, and he concludes that intentionally interfering "in humankind's genetic inheritance is not a sufficient reason to foreswear the technique forever, though it is reason enough to distinguish it from somatic-cell interventions."[24] Robert Morison's warning is an appropriate note on which to close our gene pool discussion:

> The nominalist biological position is that there can be no such thing as an ideal man. Men are brothers simply because they all draw their assortment of genes from a common pool. Each individual owes his survival and general well-being partly to his own limited assortment of characters and partly to the benefits received through cultural interchange with other individuals representing other assortments. It follows that the brothers in such a human family have a sacred obligation to maintain the richness and variety of their heritage—their human gene pool and their common culture. Every man in a sense must become his brother's keeper, but the emphasis is on keeping and expanding what both hold in common, not on converting one brother to the ideal image held by the other.[25]

Social Dangers. The intentional alteration of germline genes seems to be the real reason that some condemn it as presumptuously "playing God" and crossing a symbolic barrier beyond which medicine and mankind become involved not in treating disease, but in recreating ourselves. It is feared that we may begin to lose our footing on the slippery slope when we extend our notions of gene therapy toward "enhancement genetic engineering." This would involve insertion of a gene to enhance a specific characteristic: for example, adding an extra gene that codes for growth hormone into a normal child in an attempt to achieve a taller individual or, in the context of germline manipulation, to create taller future generations. From the medical perspective, French Anderson points out that adding a normal gene to correct the harmful effects of a nonfunctional or dysfunctional gene is different from inserting a gene to make more of an existing product. Selectively altering a characteristic might endanger the overall metabolic balance of individual cells or the body as a whole.[26] He has also warned:

> I fear that we might be like the young boy who loves to take things apart. He is bright enough to be able to disassemble a watch, and maybe even be bright enough to get it back together again so that it works. But what if he tries to "improve" it? Maybe put on bigger hands so that the watch is "better" for viewing. But if the hands are too heavy for the mechanism, the watch will run slowly, erratically, or not at all. The boy can understand what he can see, but he cannot comprehend the precise engineering calculation that determined exactly how strong each spring should be, why the gears interact in the ways that they do, etc. Attempts on his part to improve the watch will probably only harm it. We will soon be able to provide a new gene so that a given property involved in a human life would be changed—e.g., a growth hormone gene. If we were to do so simply because we could, I feel that we would be like that young boy who changed the watch hands. We, like him, do not really understand what makes the object we are tinkering with tick. Since we do not understand, we should avoid meddling. Medicine is still so inexact that any modification (except perhaps one which returns towards normal a defective property) might cause severe short-term or long-term problems.[27]

The issues raised by such enhancement efforts are similar to those of athletes taking steroids, but with the added complication of perpetuating the effects to unborn generations. But assuming that all the medical concerns can be addressed, and even if such issues as scarcity of resources and equal access to care are resolved, one major ethical problem remains—the likelihood of genetic discrimination. Using the example of inserting the gene for growth hormone, if being tall were considered a social virtue, say for basketball players, it would only be an advantage if "opponents" could be kept relatively shorter by selectively limiting their access to the "treatment." What if a gene could be inserted to prevent a certain type of cancer in susceptible, yet otherwise normal, individuals. Could they lose their children's health insurance if they refused to have their gametes or embryos "treated"?[28] Will we be able to resist "encouraging" parents to decrease the "genetic burden" and thus not undermine individual autonomy and dignity?

The most troublesome of all forms of genetic engineering is "positive" eugenics. Human beings seem to have an urge to "improve" our own species by genetic manipulations. Throughout history men and women have practiced assortive mating based on physical characteristics, intelligence, artistic talent, disposition, and many other traits. It was the English aristocrat and mathematician Francis Galton, a cousin of Charles Darwin, who in 1883 coined the term "eugenics" from the Latin word meaning "well-born." In his writings, Galton defined eugenics as the science of improving human condition through "judicious matings . . . to give the more suitable races or strains of blood a better chance of prevailing speedily over the less suitable."[29] Eugenic genetic engineering includes attempts to alter or "improve" complex human traits that are at least in part genetically determined—for example, intelligence, personality, or athletic ability. Because such traits are polygenic, purging the genome of undesirable genes and replacing them with an array of desirable genes would require technological advances that we cannot even foresee. Moreover, even if such replacement could be achieved, the interplay between the newly introduced genetic material and the recipient genome would result in entirely unpredictable results. Nonetheless, the scenario remains in the realm of remote possibility.

We need not envision a return to the "racial hygiene" totalitarianism of National Socialism under the Nazis to see that the genetic screening of pre-implantation embryos might become popular, or even standard. As the U.S. Congress's Office of Technology Assessment (OTA) put the case in 1988, such screening need not be mandated by government at all, since individuals can be made to *want* it (as they are now made to want all sorts of things by advertising), even to insist on it as their right. In the words of the OTA report, "New technologies for identifying traits and altering genes make it possible for eugenic goals to be achieved through technological as opposed to social control."[30]

Sheldon Krimsky has persuasively argued that there are two potential moral boundaries for gene therapy: the boundary between somatic cells and germline cells, and the boundary between the amelioration of disease and the enhancement of traits. But as he has noted also, the first involves a clear distinction, but a dubious rule, whereas the second involves a desirable rule, but a fuzzy distinction. The problem is that the distinction between disease and enhancement has no objective, scientific basis; disease is constantly being redefined. Krimsky asks, for example, "is chemical hypersensitivity a disease? Any trait that has a higher association with the onset of a disease may itself by typed as a proto-disease, such as fibrocystic breasts."[31]

Thus the problem is that we want to use germline gene therapy only to correct devastating diseases to avoid, among other things, the creation of a "super class" of privileged and gene-enhanced individuals who have the advantage of both wealth and enhanced genetic endowment. But the solution is unlikely to be drawing a line between disease and enhancement, both because that line is inherently fuzzy and because once "treatment" techniques are established, it may be impossible, as a practical matter, to prevent these same techniques from being used for enhancement. Because many traits one might want to enhance, such as intelligence or beauty, are polygenic, we may also comfort ourselves that they may never actually be susceptible to predictable genetic manipulation.[32]

Arguments for Human Germline Therapy

Efficiency. Leroy Walters suggested two rationales for which human germline therapy are ethically defensible.[33] The first rationale is efficiency. Assuming that somatic cell gene therapy became a successful cure for disorders caused by single-gene abnormalities, such as cystic fibrosis or sickle cell disease, treated patients would constitute a new group of phenotypically normal, homozygous "carriers" who could then transmit abnormal genes to their offspring. If a partner of such an individual had one normal copy for the gene and one abnormal one, there would be a 50 percent likelihood of an affected offspring. If two treated patients with the same genetic abnormality reproduced, all of the offspring would be affected. Each succeeding generation could be treated by means of somatic cell gene therapy; however, if available, some phenotypically cured patients would consider it more efficient, and in the long run less costly, to prevent transmission of the abnormal gene to their offspring via germline gene therapy. Andreas Gutierrez and colleagues appear to have accepted this efficiency rationale as well by suggesting that germline gene therapy might be used to *prevent* cancers in individuals carrying defective tumor suppressor genes (for example, the retinoblastoma gene in retinoblastoma and p53 in Li-Fraumeni syndrome).[34]

It has also been suggested that germline therapy would be needed to treat genetically defective embryos of couples who believe it is immoral to discard embryos (because they are human life) regardless of their genetic condition.[35] This argument, however, is not persuasive. First, individuals with this belief might not be able to justify putting their embryos at risk of extracorporeal existence in the first place, and even if they could, would find it even more difficult to justify manipulations of the embryos with germline gene therapy that may cause their demise. Further, even those who adamantly oppose abortion do not equate the failure to implant an extracorporeal human embryo with the termination of a pregnancy. Thus, germline gene therapy cannot be justified solely on the basis of the religious beliefs of those who hold that protectable human life begins at conception. Moreover, as has been argued previously, it cannot be justified solely on the basis of treating the embryos of homozygous parents, since alternatives, such as ovum and sperm donation, exist that put the potential child at no risk.

Unique Diseases. The second rationale for the germline approach would arise if some genetic diseases could only be treated by this method. For example, in hereditary diseases of the central nervous system somatic cell gene therapy may be impossible because genes could not be introduced into nerve cells due to the blood-brain barrier. Early intervention that did not distinguish between somatic cells and germ cells may be the only means available for treating cells or tissues that are not amenable to genetic repair at a later stage of development or after birth.

On the surface, the efficiency argument seems reasonable if one is dealing with a genetic characteristic in all sperm and one could remove it from all sperm by manipulaing testicular cells. On the other hand, if, as seems most likely, screening will be done on preimplantation embryos, and those with genetic "defects" identified, then the *most efficient* method of dealing with the defective embryos is to simply discard them, implanting only the "healthy" ones.

For now at least, it seems that the second rationale is the stronger one, but since we have no coherent theory for treating such diseases by germline therapy, actual experimentation is at best premature.

Should Germline Therapy Be Permitted?

Even though Jean Dausset's moratorium proposal did not pass at Valencia, there is currently a de facto moratorium on germline therapy because it is unclear both how to do it and for what conditions it might be appropriate. Consequently, a formal moratorium seems unnecessary. It also seems unwise. A review of the literature and this summary of it make it clear that the issues have not been well thought out or well debated. The arguments against germline gene therapy tend to be basically the same as those previously used against somatic cell therapy: work with genes seems to arouse greater concern primarily because of the "genetic theories" of the Nazis and their horrible acts to put them into practice, and the early work on recombinant DNA in the United States and the concern that it might create a dangerous strain of virus or an uncontrollable pathogen. The "future generation" argument is actually no different than the original arguments raised against IVF and any extracorporeal manipulation of the human embryo.

What seems most reasonable now is to continue the public debate on whether and under what conditions germline experimentation should be attempted. As a way to better focus this debate, we recommend the following prerequisites be met prior to attempting any human germline gene therapy:

1. Germline gene experimentation should only be undertaken to correct serious genetic disorders (for example, Tay Sachs disease).
2. There should be considerable prior experience with human somatic cell gene therapy, which has clearly established its safety and efficacy.

3. There should be reasonable scientific evidence using appropriate animal models that germline gene therapy will cure or prevent the disease in question and not cause any harm.
4. Interventions should be undertaken only with the informed, voluntary, competent, and understanding consent of all individuals involved.
5. In addition to approval by expert panels such as the NIH's Working Group on Gene Therapy and local Institutional Review Boards, all proposals should have prior public discussion.

An international consensus is desirable, because germline gene manipulation is the area in which there is the most international concern. This presents us with the first real opportunity to develop an international forum for policy debate and perhaps even resolution. Since we are dealing with the future of the species, this does not seem like too much to expect.

NOTES

1. D. J. Weatherall, "Gene Therapy in Perspective," *Nature* 349:275–76, 1991; B. J. Culliton, "Gene Therapy on the Move," *Nature* 354–429, 1991; and M. Hoffman, "Putting New Muscle into Gene Therapy," *Science* 254:1455–56, 1991.
2. S. A. Rosenberg, P. Aebersold, K. Cornetta, A. Kasid, R. A. Morgan, R. Moen, E. M. Karson, M. T. Lotze, J. C. Yang, S. L. Topalian, M. J. Merino, K. Culver, A. D. Miller, R. M. Blaese, and W. F. Anderson, "Gene Transfer into Humans— Immunotherapy of Patients with Advanced Melanomas, Using Tumor-Infiltrating Lymphocytes Modified by Retroviral Gene Transduction," *New England Journal of Medicine* 323:570–8, 1990.
3. A. A. Gutierrez, N. R. Lemoine, and K. Sikora, "Gene therapy for cancer," *Lancet* 339:715–21, 1992; and A. Abbott, "Italians First to Use Stem Cells," *Nature* 356:465, 1992.
4. L. Thompson, "Human Gene Therapy Debuts at NIH," *Washington Post,* September 15, 1990, p. A1. Other researchers may claim to have conducted the first human gene transfer experiments. In the early 1970s, German researchers conducted an experiment on German sisters who suffered from a rare metabolic error that caused them to develop high blood levels of arginine. Left uncorrected, this genetic defect leads to metabolic abnormalities and mental retardation. Using Shope virus (which induces a low level of arginine in exposed humans), the researchers infected the girls in the hope that the virus would transfer its gene for the enzyme that the body needs to metabolize arginine. The attempt failed. The next experiments took place in 1980 in Italy and Israel. Turned down by the UCLA Institutional Review Board for an experiment to introduce the globin gene (by mixing the patient's bone marrow with cells with DNA coding for hemoglobin in the hope that a normal hemoglobin gene would stably incorporate into the bone marrow cells) in a patient with beta-thalassemia, Dr. Martin Cline later unsuc-

cessfully performed this experiment on two children, one in Italy and one in Israel. He was sanctioned by the NIH for failure to obtain IRB approval and his case became notorious. See President's Commission for the Study of Ethical Problems in Medicine and Biomedical and Behavioral Research, *Splicing Life,* U.S. Government Printing Office, Washington, D.C., 1982, pp. 44–5; and materials in J. Areen, P. King, S. Goldberg, and A. M. Capron, *Law, Science and Medicine,* Foundation Press, Mineola, NY, 1984, pp. 165–70.

5. For example, "Gene Therapy (Editorial)," *Lancet* 1:193–4, 1989; G. Kolata, "Why Gene Therapy Is Considered Scary but Cell Therapy Isn't," *New York Times,* September 16, 1990, p. E5.

6. E. K. Nichols, *Human Gene Therapy,* Institute of Medicine, National Academy of Science, Harvard Press, Cambridge, MA, 1988, p. 163.

7. President's Commission for the Study of Ethical Problems in Medicine and Biomedical and Behavioral Research, *Splicing Life,* U.S. Government Printing Office, Stock no. 83-600500, Washington, D.C., 1982.

8. "Gene Therapy in Man. Recommendations of European Research Councils," *Lancet* 1:1271–2, 1988.

9. W. F. Anderson, "Human Gene Therapy: Scientific and Ethical Considerations," *Journal of Medical Philosophy* 10:275–91, 1985.

10. B. Culliton, "Gene Therapy: Into the Home Stretch," *Science* 249:974–6, 1990.

11. Department of Health and Human Services, "National Institutes of Health Points to Consider in the Design and Submission of Human Somatic-Cell Gene Therapy Protocols," *Recombinant DNA Technical Bulletin* 9:221–42, 1986.

12. Ibid.

13. W. F. Anderson, "Human Gene Therapy: Why Draw a Line?" *Journal of Medical Philosophy* 14:681, 1989.

14. Ibid.

15. W. F. Anderson, "What's the Rush?" *Human Gene Therapy* 1:109–10, 1990.

16. U.S. Congress, Office of Technology Assessment, New Developments in Biotechnology—Background Paper, *Public Perceptions of Biotechnology,* OTA-BBP-BA-45, U.S. Government Printing Office, Washington, D.C., May 1987.

17. G. Fowler, E. T. Juengst, and B. K. Zimmerman, "Germ-Line Gene Therapy and the Clinical Ethos of Medical Genetics," *Theoretical Medicine* 10:151–65, 1989.

18. S. Elias and G. J. Annas, *Reproductive Genetics and the Law,* Year Book, New York, 1987.

19. L. Walters, "The Ethics of Human Gene Therapy," *Nature* 320:225–7, 1986.

20. Fowler et al., 1989, and Nichols, 1988. And see generally, on human embryo experiments, P. Singer and H. Kuhse, "The Ethics of Embryo Research," and G. J. Annas "The Ethics of Embryo Research: Not as Easy as It Sounds," *Law, Medicine and Health Care* 14:133–40, 1987.

21. Anderson, 1985.

22. A. G. Motulsky, "Impact of Genetic Manipulation on Society and Medicine," *Science* 219:135–40, 1983.

23. A. Capron, "Which Ills to Bear: Reevaluating the 'Threat' of Modern Genetics," *Emory Law Journal* 29:665–96, 1990.

24. Ibid.

25. R. Morison, "Darwinism: Foundation for an Ethical System?" *Zygon* 1:352, 1966.

26. Anderson, 1985, 1989.
27. W. F. Anderson, "Human Gene Therapy: Where to Draw the Line," unpublished draft, 1986, p. 607, reprinted in 1987 supplement to Areen et al., 1984, suppl., p. 27.
28. Anderson, 1985, 1989.
29. D. Susuki and P. Knudtson, *Genetics: The Clash between the New Genetics and Human Values,* Harvard Press, Cambridge, MA, 1989.
30. U.S. Congress, Office of Technology Assessment, *Mapping Our Genes,* OTA-BA-373, U.S. Government Printing Office, Washington, D.C., April 1988.
31. S. Krimsky, "Human Gene Therapy: Must We Know Where to Stop before We Start?" *Human Gene Therapy* 1:171–3, 1990.
32. B. Davis, "Limits to Genetic Intervention in Humans: Somatic and Germline," in *Human Genetic Information: Science, Law and Ethics,* Ciba Foundation Symposium 149, John Wiley and Sons, New York, 1990.
33. Walters, 1986.
34. Gutierrez, 1992.
35. R. M. Cook-Deegan, "Human Gene Therapy and Congress," *Human Gene Therapy* 1:163–70, 1990.

IV

How Changes in Genetics Change Clinical Practice

9

Privacy and Control of Genetic Information

Ruth Macklin

In Western societies in which liberal individualism is the predominant ideology, the importance of privacy as an ethical value is generally assumed. Like individual liberty, personal privacy occupies a high position in a hierarchy of values, and actual or proposed intrusions normally require justification. However, personal privacy is not a universally recognized value. When Victor Sidel, a physician well known for his national and international work in public health, visited the People's Republic of China some years ago, he encountered common public-health practices that required people to reveal highly personal information, which was then posted in a public place. Sidel asked, "Don't people consider this an invasion of their privacy?" and his Chinese interpreter could not translate the question. The Chinese language apparently lacked a concept of privacy in the sense that makes it an ethical value in Western society.

The fact that privacy is not a universally cherished value does not mean that no objective grounds can be given for holding it to be important. Even in cultures whose members do not feel wronged when their privacy is invaded, people's interests may still be harmed when personal information about them is revealed to others. The objective grounds for respecting privacy is the potential for harm to individuals whose privacy is invaded. Where the culture as a whole respects the value of privacy, people are wronged by intrusions of privacy even if no harm befalls them as a result of the intrusion. Since Western democracies generally hold privacy to be an ethical value, invasions of privacy can both wrong people and stand to harm their interests when confidentiality is breached.

Why Genetic Information Is Private

It is evident that genetic information is personal information, in a morally neutral sense of the term "personal." An individual's height, skin color, and observable physical characteristics are also personal in this neutral sense. They are personal because they are about the individual. At the same time, these physical characteristics are visible to any casual observer. An individual's birth date is also personal information, often taken to be private as well as personal. People can conceal their birth date, and some—men as well as women—take elaborate steps to conceal their true age. Yet an individual's date of birth may justifiably be required for certain purposes, for example, obtaining a driver's license or buying alcohol. What these observations show is that not all personal information can be considered private, and personal information is often held to be private in some contexts or for some purposes, but not for others.

Genetic information falls into the more general category of medical information, which is protected by a long-standing tradition of privacy and confidentiality in the physician-patient relationship. Edgar and Sandomire note that "[m]edical privacy . . . is valued for itself and is valued for the protection of other interests its waiver may implicate."[1] The role of physicians grants them moral license to inquire about their patients' private information and prohibits them from disclosing that information except in certain circumstances, and then only to authorized individuals or agencies. To provide proper health care, professionals must often know the most intimate details about the habits or actions of patients. It is unlikely that patients would disclose such details unless they are confident that only people directly involved in their care will learn of them.[2]

The ethical and legal presumption is clearly on the side of preserving confidentiality of information that relates to patients' and clients' diagnoses, prognoses, and other facts. Yet in the hospital setting, it is not realistic for patients to expect that information about them will remain confidential.[3] Medical records may be viewed by many individuals in the hospital, not restricted to other health professionals, but also including administrators, secretaries, and financial officers. George Annas urges that "information access should be limited to a 'need-to-know' basis," but cautions that "the reality is that many people will fit into this category and thus have free access to the patient's record."[4] For medical care delivered outside the hospital, in a private physician's office or in an outpatient setting in a health facility, it should be possible to provide better protection for confidential information about patients.

The reasons for having ethical principles and laws protecting physician-patient confidentiality are several. First, clinical information may be extremely sensitive, and disclosure to unauthorized persons has the potential to harm a patient's interests. To take a recent and, perhaps, extreme example, there is ample evidence that the interests of people with AIDS have been seriously harmed by disclosure of their HIV infection: loss of housing, jobs, and insurance are among the documented consequences of disclosure for many individuals. Even when the consequences of disclosure are not likely to be so damaging, people would feel embarrassed if intimate personal details could readily be revealed.

The second reason for ethical and legal protection is that the physician-patient relationship is built on trust. When physicians breach confidentiality, they undermine trust and damage the physician-patient relationship, and this might interfere with the healing process. It is especially important to maintain trust when counseling is an integral part of the physician-patient relationship.

Genetic information is private as well as personal. There is a subjective as well as an objective basis for this. In a culture that values privacy, it is the kind of information people have come to treat as private and expect their doctors not to reveal. Patients do not feel intruded upon when their physician asks them the most intimate questions. Patients assume that when their physician requests information, it is to be used for their own benefit. That is the subjective aspect.

The objective aspect derives from the physician's role and the purpose for which private information is sought. Doctors and patients have an identity of interests: the patient's health and well-being. However, when others outside the physician-patient relationship seek information, it cannot readily be assumed that it is obtained for the sole or primary purpose of benefitting that individual. We have only to think of the usual suspects—employers and insurance companies—who might be eager to obtain genetic information for a purpose other than that of benefitting the person to whom the information applies. If genetic information can properly be considered private, then outside the physician-patient relationship, requests or demands for such information from the individual to whom it applies constitute unwarranted invasions of privacy.

A feature of information derived from mapping the human genome that goes beyond the ordinary varieties of medical information is its predictive value. It holds the prospect for transforming suspicions about genetic susceptibilities into probabilistic statements about the likelihood of a given individual contracting cancer, hypertension, or Alzheimer's disease. If information about a person's past medical history or current condition should be considered private, it is all the more evident why information pointing to the individual's future prospects deserves that moral status. Well-documented evidence of harm that has befallen people's interests as a result of disclosure of their past psychiatric history or their current HIV disease, to name only two examples, are more than adequate to illustrate the legitimacy of such worries.

The claim that genetic information should be treated as private has obvious implications for the control of such information. It ought to remain in the hands of the person to whom it applies and revealed to others only with that individual's consent. Whether this ethical conclusion will be respected by policy makers in the future is something about which we can only speculate.

The "Need to Know"

To observe that private information is sometimes used in a manner that harms people's interests does not resolve controversies about when such intrusions of privacy or breaches of confidentiality may be justified. Even the staunchest defenders of individual privacy and protecting confidentiality admit some circumstances in which those values may justifiably be overridden. A high likelihood of harm to others can justify placing limits on individual liberty. Similar circumstances could justify placing limits on respect for privacy and confidentiality. Although the general ethical and legal principle is that "medical tests or diagnoses shall not be indiscriminately revealed, [they may be divulged] when confidentiality is outweighed by countervailing concerns."[5] These overriding considerations are often couched in terms of the "need to know."

The so-called need to know is one of the more elusive concepts to find its way into laws and institutional policies. In contrast to a want or a desire, a *need* is something that must be fulfilled. We have only to think of the moral claims that arise out of human needs, for example the need for food, shelter, or medical supplies, and it is immediately evident that calling something a need creates a presumption in favor of its fulfillment. However, it is difficult to distinguish the need to know certain information from the mere desire or wish to know.

An example can be found in the New York State AIDS legislation. The law is designed, in large part, to protect the confidentiality of HIV information and to prevent unauthorized disclosure of such information. However, the law specifies a number of exceptions to the prohibition against disclosing HIV information. Among the individuals or agencies to whom AIDS-related information may be disclosed without the consent of "the protected individual" (as HIV-positive people are termed in the legislation) are health care providers and facilities, insurers and other third-party payors, state and local health officers, and various governmental agencies, including adoption, foster care, parole, probation, and those that regulate health care providers.[6]

Disclosure to health professionals and institutions could arguably serve either of two purposes: the HIV-infected person's interest in receiving appropriate medical treatment, or the interests of other individuals in protection from infection.[7] Considerable controversy surrounds the latter purpose, and debates about the need to know patients' HIV status continue to rage in hospitals and among health professionals. In contrast to both these purposes, however, is the provision in the New York State legislation allowing disclosure to insurance companies in order to reimburse physicians and others for provision of HIV-related services.

On the consent form signed by people whose blood is tested for HIV, the list of these individuals or agencies to whom HIV information may be disclosed without the individual's consent mentions "others on a need to know basis." Referring to the confidentiality protections in the New York State AIDS statute, one author remarks that "there is some danger . . . that the exceptions . . . may swallow the rule."[8]

There is no objective or agreed-upon definition of "need to know" in reference to information about individuals that is essentially private. It would not be surprising if legislation that pertains to information obtained from the Human Genome Project will reflect the desire on the part of people or agencies to know, rather than their genuine need.

An enlightened discussion of the need to know in a Model HIV Confidentiality Policy[9] proposes that the concept be interpreted in a way that serves the needs of clients, not those of the staff member or anyone else. The discussion notes that different individuals are likely to have a different idea about what qualifies as a need to know HIV-related information. "Is 'need' defined from the client's perspective or the staff member's? Some policies refer to a 'legitimate need to know' but this begs the question of what makes a particular need legitimate."[10] Although the analogy between HIV-related information and genetic information is not perfect, lessons from the ongoing AIDS epidemic are nevertheless instructive.

Another lesson demonstrates the gap between protections designed to prevent intrusions of privacy and the ease of disclosure once that same information has been gathered. In New York State, the law has so far prohibited testing infants for HIV without the consent of the child's biological parent or legal guardian. A prospective foster parent could not demand that a child be tested prior to placement in the home. However, once a child has been tested for HIV, the information may be revealed to prospective foster parents in advance of placement.

A "Duty to Disclose"?

A concept invoked more frequently than the need to know medical information normally treated as confidential is that a "duty to disclose" such information. A well-known legal precedent regarding this obligation of therapists was set in the Tarasoff case, decided in California in 1976. It is worth summarizing this landmark case briefly.[11]

Tatiana Tarasoff was a student at the University of California at Berkeley who was killed by Prosenjit Poddar, another student who met her in a folk-dancing class and fell in love with her. Tarasoff rebuffed Poddar, who fell into a severe depression and after six months sought psychiatric help as an out-patient at the university health service. The therapist, a clinical psychologist, found Poddar "a danger to the welfare of other people and himself" after he had disclosed his intention to kill Tatiana Tarasoff when she returned from a summer in Brazil. The psychotherapist contacted the campus police, who detained Poddar but then released him when he seemed rational and made a promise to stay away from Tarasoff. However, when she returned from Brazil two months later, Poddar killed her with a butcher knife. Neither the victim nor her parents had been warned of the threat. The parents brought a wrongful death action against the Board of Regents of the State of California, the campus police, and the doctors who had been involved in Poddar's treatment.

The California Supreme Court ruling in the case imposed on psychotherapists a duty to warn a potential victim, or to take other preventive steps, if there is reason to believe that person is in danger from the therapist's patient. At the time of the court decision, the *Tarasoff* case gave rise to considerable concern because of the special importance of confidentiality in the psychotherapist-patient encounter: it is held to be vital to the success of the therapy itself.

A dissenting opinion in *Tarasoff* argued on the basis of two different considerations. The first is the damage done to the therapist-patient relationship, damage that may lead fewer people to seek treatment in the first place and will inhibit those who do from making the full disclosure necessary for effective treatment. The second consideration is the limited ability of the psychiatric professional to predict dangerousness.

Although *Tarasoff* involved a psychiatric patient, setting a precedent for the obligation of therapists to do something to protect specific persons from being harmed by an individual judged to be violent or dangerous, the court in *Tarasoff* partially "based its decision in part on its reading of earlier cases holding physicians liable for failing to inform family members of the infectious nature of a patient's condition."[12] The early cases are those of persons with infectious diseases, rather than those likely to harm others by violent behavior.

The principal reason for allowable breach centers on harm to others that might come from *not* breaching confidentiality. But this only invites a series of subsidiary questions, which should be addressed any time a deliberate breach of confidentiality is contemplated. The chief questions are: How high should the probability be that harm will occur? On what is the assessment of probability based? Do experts agree on that assessment? How serious must the probable harm be? Are there alternative means for seeking to prevent it besides breaching confidentiality? To whom may disclosure be made? To authorities, such as the police? To other health care professionals or administrators? To the patient's family? To an employer? Must the patient be informed that the physician intends to disclose confidential information? If so, must the patient be informed prior to disclosure? Or is it sufficient to notify the patient after the disclosure has been made? Will disclosure of confidential information actually have the intended effect, whatever that effect is?

If the answers to these questions can be truthfully and satisfactorily answered each time a breach of confidentiality is contemplated, the instances in which an actual duty to disclose medical information emerges will be few. In some situations involving highly contagious diseases or emotionally disturbed individuals evaluated as dangerous to others, the probability and magnitude of harm likely to befall people who are not informed may be considerable. In the case of genetic information, however, occasions in which harms stemming from nondisclosure can be identified and reliably predicted are likely to be quite rare. The chief category appears to be that of genetic information that can substantially affect a patient's relatives. If such information is disclosed, it can enable them to make reproductive decisions and other life plans in accordance with the newly revealed information.

Without adequate information, people cannot make informed choices and decisions related to their own health and well-being. If a physician or counselor learns genetic information about a patient that could substantially affect the interests of the patient's relatives, should that give rise to a *duty* to disclose? This situation constitutes a genuine dilemma in which the physician or counselor has two prima facie obligations, both of which cannot simultaneously be met: a duty to the patient to preserve confidentiality, versus a duty to disclose medically relevant information to a third party who could benefit substantially from that information. An initial effort should obviously be made to convince the patient to do the disclosing, or at least to grant permission for the physician to reveal information to the relatives. If the patient refuses, where does the professional's greater obligation lie?

There can be no general answer to this question, other than a reminder of the presumption favoring confidentiality. However, once an ethical analysis leaves the realm of rights and moves to an assessment of the consequences of actions, it becomes a delicate exercise in balancing the risks and harms of nondisclosure against those of disclosure. As is true of any such balancing attempts, the decision maker should strive for an honest assessment of the good and bad consequences, and not rest content with assumptions that flow from preconceived value preferences.

If a physician discovers that a patient has a genetic disorder that could increase the chances of morbidity and mortality for a similarly diseased relative, the physician may disclose information the relative could use in seeking to minimize the likelihood of the worst health outcomes. The situation is ethically murkier when a genetic disease has no cure or effective treatment, and the only purpose that could be served by disclosure is to inform the unwitting relative about the prospect of bad things to come. Because it is not clear that disclosure would truly benefit the relative, the physician may honor the patient's wish to preserve confidentiality.

Nevertheless, there are some circumstances, albeit few, in which physicians or counselors are ethically justified in breaching confidentiality for the purpose of disclosing beneficial information to a patient's relatives. This suggests that those who screen or test for genetic conditions should inform patients or clients of that possibility in advance. A sort of "genetic Miranda warning" could allow a patient who insists on strict confidentiality to act accordingly: avoid the test altogether, seek another physician or clinic, or be forewarned that private information may end up being revealed. To inform patients in advance that some circumstances may arise in which genetic information will be disclosed is not likely to jeopardize the physician-patient relationship, so long as the physician assures the patient that the presumption lies in favor of confidentiality and outlines the exceptions to that presumption.

It is important to note that any ethically permissible or obligatory disclosure of this sort would be limited to the individuals who might directly and significantly be affected by learning the information. This is a very different situation from breaches of confidentiality stemming from access to storage of sensitive personal information. Access to databases and the potential for abuse of stored information raise a different, more troubling prospect in an era of computerized data banking.

Escalation of Data Banking

The type and amount of information likely to be produced by the Human Genome Project are only one cause for concern about the control of genetic information. Worries about control might be limited were it not for another major technological revolution: the creation and spread of national and international networks of computerized information resources. In addition to specific data sources, such as credit rating services and other data banks that store a specific, limited set of data about individuals, there is also the phenomenon of "statistical disclosure," which refers to the ability of someone with access to a number of different databases to pose a series of queries, requiring only yes-or-no answers, which can yield highly specific information about an individual.

There is already a substantial legislative interest in creating DNA data banks, although these are mostly limited at present to convicted felons. In a report prepared by the Technical Working Group on DNA Analysis Methods (TWGDAM), the FBI has proposed creating a national DNA database for the purpose of law enforcement. A statement of the initial plan asserts that "the database will be limited to profiles of persons convicted of violent crimes or those involved in unsolved cases. No names of individuals or other identifying or characterizing data will be stored with the DNA profiles."[13]

It should be clear that this type of DNA information is limited to DNA profiles, which provide only limited information about an individual. The proposed storage of DNA profiles in state or national data banks for law enforcement purposes contains information about individual identity only. Nevertheless, a report by the Office of Technology Assessment (OTA) on forensic uses of DNA tests notes that "because DNA is specific to an individual and so highly personal, some are reluctant to see any DNA test results become part of a de facto national database."[14] The OTA report also mentions the concern some people have about the slippery slope: the "fear that genetic testing will not be limited to identity, but will expand to include disease (e.g., sickle cell or Huntington disease), proclivity toward disease (e.g., cancer or coronary disease), or behavioral characteristics (e.g., schizophrenia) that could then find their way into the database."[15]

The TWGDAM report of the FBI proposal addresses these concerns under the heading "Privacy Concerns and Data Security Issues." The report offers assurance that "there is no known relationship between these numbers and any physical or mental condition."[16] In addition, the report states that "probes will not be used that are linked to disease conditions or personality traits. Only information relevant to law enforcement requirements will be maintained in the national DNA database."[17]

On the question of access to the national DNA database, the report has this to say: "Access to the national DNA database will be limited to government agencies on a 'need to know' basis as described below. Access by private contractors and the military, in the service of the federal or state government . . . , would be through the appropriate law enforcement agency only."[18] The assurance that access will be limited to government agencies on a need to know basis provides little comfort, especially to those familiar with the FBI's surveillance of citizens for political purposes over the years.

A recent article in the *New York Times* described the FBI files on the J. Roderick MacArthur Foundation and its president. The foundation has supported the defense of human rights in El Salvador and Guatemala, among other countries "with egregiously repressive governments that are heavily supported by the U.S.,"[19] and is now suing the FBI to find out why it is keeping files on the family Foundation. The author of the article, a member of the Foundation's board, explains the FBI's interest in the Foundation by noting that the agency "still views opposition to Government policy as subversive and dangerous, particularly when it involves Latin America and Southeast Asia."[20] The article urges the U.S. Congress to "banish once and for all the ghost of Hoover. . . . After more than 40 years as a quasi-political force, the Bureau should devote its full attention to its original purpose: catching criminals."[21]

The FBI's TWGDAM report proposes inclusion of the following types of indices in the national database: a statistical DNA population database, devoid of any personal identifiers; open-case DNA profiles obtained from body fluid stains found on evidentiary materials recovered from violent crime cases having no suspects; convicted violent offender DNA profiles; and missing persons/unidentified bodies DNA profiles.[22] The justification offered for the development of these types of indices "is based on the fact that individuals who commit violent crimes are often repeat offenders."[23] This would make it possible to link sexual assault cases together, thus enabling the tracking of serial rapists. "In light of the high recidivism rate (50%), the convicted sex offender data would provide the investigator with a logical first place to look for assistance in solving unknown subject sexual assault cases."[24]

Unlike the questions addressed earlier regarding physicians' acquisition and disclosure of private genetic information, the involvement of the government in data banking raises the issue of constitutional rights. It does not take much imagination or wild extrapolation to prompt concerns about where national data banking that begins with storing DNA profiles of people convicted of violent crimes might lead.

We find ourselves once again contemplating the infamous slippery slope. The usual caveats about the slope must be issued, of course: doubts about the likelihood that a slide is imminent, or about the difficulty of stopping a slide once it has begun. Having issued the caveat, let me turn to the slope.

The Slippery Slope

As any skier knows, there is always more than one slope on a mountain. Some may be more slippery than others. At least the following considerations are worthy of attention.

Perhaps only a cynical view of the FBI and its practices provides cause for worry about its proposal to create a national felon database. The TWGDAM report emphasizes that the purpose is to support law enforcement operations only. But in describing the components of the database system, the report states that "the investigative support database would *initially* include open case, convicted violent offender and missing person/unidentified body DNA profiles" (emphasis added).[25] It is only natural to wonder what it might *eventually* include.

The state of Virginia is a leader in DNA testing for forensic purposes. In 1989, legislation was enacted that called for mandatory samples from all convicted sex offenders.[26] Scientists at a state DNA typing laboratory were given training both by Lifecodes Corp., a private company, and by the FBI. Recently, the Virginia Bureau of Forensic Science decided to switch from the DNA testing system used by Lifecodes to that used by the FBI, to promote standardization that would be important for a national data bank.[27]

Although in Virginia no agencies other than those involved in criminal justice plan to use the state DNA typing facility, broader uses have been proposed in other states.[28] A resolution proposing the use of uniform technology in developing DNA profiling indices was adopted at the December 1989 meeting of the National Association of Attorneys General. Among other items, the resolution encourages "the states to develop DNA profiling indices, *including, but not limited to, individuals convicted of violent crimes* and evidence recovered from the scenes of violent crimes" (emphasis added).[29] These developments lead to speculation about the evolution of large data banks that begin with creating felon databases and eventually encompass other categories. A jump has already occurred from maintaining profiles of convicted sex offenders, to those convicted of other violent crimes, to individuals convicted of misdemeanors. Informal speculation surmises that the next wave of data banking will occur in the military, for newborns (storing blood samples from all newborns for later analysis), and for use by organ transplantation systems.

The worries just noted about the slippery slope refer only to the creation of data banks containing DNA profiles used for identification. A connecting slope prompts the question: How certain can we be that the genetic information stored in a national data bank will be limited to DNA profiles used for identification purposes? A cautionary tale serves as a reminder that even the medical profession has been complicit in the FBI's agenda.

In February 1972, the *Archives of Dermatology* contained an FBI "wanted" poster on its pages. The notice, which also appeared in the *Archives of Internal Medicine,* described the woman pictured in the FBI poster as afflicted with an "acute and recurrent" skin condition that "could necessitate treatment by a dermatologist."[30] The thirty-year-old woman had been indicted by a grand jury for "conspiring with another individual" in an act involving the interstate transportation of explosives. The alleged violation had occurred two years earlier and no trial had yet been conducted. Both journals are official publications of the American Medical Association, whose spokespersons at the time denied that there were any questions of medical ethics involved in publishing FBI wanted posters in medical journals.[31]

When a data bank contains actual stored samples, rather than DNA information profiles derived from samples, a much greater potential exists for extracting information beyond the purpose for which the sample was originally collected. An article on DNA data banks observes that "until the government decides upon a uniform method of DNA analysis and appropriate standards, laboratories preserve the actual bodily fluid in a deep freeze for future testing."[32] A California statute does not expressly limit how stored blood and saliva samples taken from convicted sex offenders may be used, although laws in other places specifically limit the purpose to identification.[33] The point here is that if actual DNA samples are stored, it leaves open the possibility of violating restrictive laws to gather more information about an individual than the laws currently allow. Moreover, it is always possible for laws to be amended to serve other law-enforcement purposes or entirely different goals.

A case in point is the prospect for universal newborn data banking. A private laboratory in New Jersey already offers a service in which it stores an infant's DNA pattern. The purpose is to establish a registry that can enable identification in case a child is lost or kidnapped. The DNA pattern is kept until the child becomes eighteen, when he or she can decide to destroy the profile or pay to keep it in the registry.[34] This development appears harmless enough, because the data banking is voluntary (at least on the part of parents) and the purpose seems reasonable. But if the program is expanded to include the storage of blood samples, and if it is carried out on a "routine" basis by legislative mandate, the implications are vast.

To envision only one possible implication, consider the practice of phenylketonuria (PKU) screening. Over the past twenty years, following the passage of laws mandating newborn screening for PKU, every state that required PKU screening obtained blood samples for that purpose. It would be a simple matter to expand the use of those samples to include extracting DNA information that could be used for other purposes deemed medically beneficial. Safeguards would be needed to prevent the unauthorized use of stored samples. The right to have stored blood samples destroyed could be assured to individuals in the same way guaranteed by the private laboratory in New Jersey for individuals whose parents stored their DNA identification samples at birth.

However, it is often hard to predict just which medical advances will occur and how beneficial they may be viewed by the medical community. Here again, an analogy with HIV testing is instructive. Earlier in the AIDS epidemic, when it became possible to test blood for HIV antibodies but when little could be done to treat the disease, there was general agreement that newborns should not be tested without their mothers' permission. Since a positive HIV test in an infant identifies the mother as positive, to allow universal newborn testing would amount to testing the women themselves without their consent. Leaving aside the zealots who have called for universal screening or testing all hospital admissions for HIV, physicians and civil liberties advocates were initially allied in their opposition to testing infants without their mothers' consent.

Now, however, that alliance has broken apart. Many physicians, pediatricians in particular, argue that early intervention with drugs is beneficial to infants who are HIV positive, because some infections can be prevented or their severity reduced. Despite the fact that about 70 percent of those who initially test positive will seroconvert to negative, pediatricians argue that the benefits to the few (the minority of infants who will not seroconvert to negative) justify testing all infants for HIV and treating those who test positive.

The point here is not to join that controversy over HIV testing of all newborns or seek to resolve it. Rather, it is to detect another slippery slope around the mountain and explore how best to negotiate it. This is the prospect that information yielded by the Human Genome Project will be seen by the medical profession as potentially beneficial to infants and children, thereby providing justification for collecting and storing a large body of genetic information. If universal storage of genetic information can lead to beneficial gene therapy, which would succeed in preventing future afflictions, pediatricians will encourage it. That may very well be a good thing for children and the adults they will grow up to be. But will the price to be paid be a steady erosion of the privacy of genetic information and loss of our control over it? Would that be a reasonable price to pay for enhancing benefits to children by providing genetic therapy?

Conclusions

To overturn existing ethical and legal presumptions in favor of protecting individuals' rights to privacy and confidentiality, a significant benefit to specific individuals or to the public must be identified. For the government to create DNA data banks and routinely store information yielded by the Human Genome Project, a compelling state interest would have to be demonstrated. Even if genetic therapy advances to the point where real benefits to genetically diseased individuals could be achieved, it is stretching the traditional interpretation of the state's public health obligation to elevate those benefits to the level of a "compelling state interest." Although it can be argued that there is a state interest in promoting and protecting the health of the population, that interest is best understood in the context of preventing the spread of infectious diseases.

The state's interest in a healthy population has not so far given rise to a recognized right to health care in the United States. Even if that shameful policy omission were to be rectified by passage of national health insurance legislation, it could not reasonably be interpreted as creating a duty on the part of citizens to undergo screening, testing, or therapy of any sort. Therefore, no basis exists, now or in the foreseeable future, for establishing national genetic data banks on the grounds that their absence would pose a risk of harm to the general public.

The possibility remains, however, for individuals to elect voluntarily to have genetic information stored in private or public data banks. If and when such systems are put in place, they should be accompanied by legal safeguards to ensure that information voluntarily entered into such data banks can equally voluntarily be removed. People who elect to undergo genetic testing or storage of their DNA samples must be informed in advance whether anyone else will have access to their DNA information, as well as how to remove the information or have the samples destroyed.

As for genetic information left in the hands of physicians and genetic counselors, circumstances can arise in which relatives could suffer harm if pertinent genetic information were not revealed to them. If a physician is unable to persuade the patient that a relative could receive substantial benefit from knowing genetic information, it is ethically permissible for the physician to inform the relative. This is analogous to situations in which it is permissible to disclose confidential information to people at risk of becoming infected with a serious disease or of being harmed by a patient deemed dangerous to others. The ethical justification for breaching confidentiality lies in the preponderance of significant harms that would likely result from adhering to the usual dictates of confidentiality.

The Human Genome Project holds the promise of furnishing people with increased benefits through genetic counseling, and in the more distant future, therapy for genetic disorders. Those benefits will best be realized if they come about without accompanying losses in the privacy of genetic information and individuals' loss of control over it.

NOTES

1. Harold Edgar and Hazel Sandomire, "Medical Privacy Issues in the Age of AIDS: Legislative Options," *American Journal of Law and Medicine*, 16:160, 1990.
2. George J. Annas, *The Rights of Patients*, 2nd ed., Southern Illinois University Press, Carbondale, IL, 1989, p. 177.
3. Ibid., p. 178.
4. Ibid. See discussion on the "need to know."
5. Edgar and Sandomire, 1990, p. 163.
6. Martha A. Field, "Testing for AIDS: Uses and Abuses," *American Journal of Law and Medicine* 16:49, 1990.
7. See Edgar and Sandomire, 1990, p. 164ff.
8. Field, 1990, p. 49.
9. Sharon Rennert, *AIDS/HIV and Confidentiality: Model Policy and Procedures*, American Bar Association, Washington, D.C., 1991.
10. Ibid., p. 45.
11. The discussion of the Tarasoff case is adapted from the discussion in Ruth Macklin, "HIV-Infected Psychiatric Patients: Beyond Confidentiality," *Ethics and Behavior*, 1:3–20, 1991.
12. Paul S. Appelbaum, "AIDS, Psychiatry, and the Law," *Hospital and Community Psychiatry* 39:14, 1988.
13. Technical Working Group on DNA Analysis Methods (TWGDAM), "The Combined DNA Index System (CODIS): A Theoretical Model," FBI Laboratory, Quantico, Virginia, October 15, 1989, p. 284; and see National Research Council, *DNA Technology in Forensic Science*, National Academy Press, Washington, D.C., 1992.
14. U.S. Congress, Office of Technology Assessment, *Genetic Witness: Forensic Uses of DNA Tests*, OTA-BA-438 U.S. Government Printing Office, Washington, D.C., July 1990, p. 21.
15. Ibid.
16. TWGDAM Report, p. 284.
17. Ibid.
18. Ibid.
19. John R. MacArthur, "Chilled by Hoover's Ghost," *New York Times*, December 6, 1990.
20. Ibid.
21. Ibid.
22. TWGDAM report, pp. 284–6.

23. Ibid., p. 286.
24. Ibid.
25. Ibid., p. 284.
26. OTA report, 1990, p. 151.
27. Ibid.
28. Ibid., p. 150.
29. National Association of Attorneys General, Resolution, adopted at the Winter Meeting, December 10–13, 1989, Phoenix, AZ.
30. Willard Gaylin, "What's an FBI Poster Doing in a Nice Journal like That?" reprinted in *Moral Problems in Medicine,* 2nd ed., edited by Samuel Gorovitz et al., Prentice-Hall, Englewood Cliffs, NJ, 1983, p. 251.
31. Ibid., p. 252.
32. Andrea de Gorgey, "The Advent of DNA Databanks: Implications for Information Privacy," *American Journal of Law and Medicine* 16:388, 1990.
33. Ibid.
34. Ibid., p. 389, note 46.

10

Genetic Predisposition and the Human Genome Project: Case Illustrations of Clinical Problems

Part I: Ray White
Part II: C. Thomas Caskey

I. Genetics of Colon Cancer Predisposition

Ray White

Human geneticists have a problem. Finally, after years of effort, we are beginning to resolve and identify the genetic components of a number of genetically transmitted disorders and predispositions. On the eve of this scientific triumph, however, at a time when we should be delivering this new knowledge to affected individuals, we have instead discovered that this delivery is compromised by social, economic, and ethical issues. Before launching into the specific example of colon cancer predisposition, which is the major problem of my own group these days, a brief discussion of the difference between a genetic disease and a genetic predisposition may be useful.

The functional distinction between an inherited disease and a predisposition lies in the probability that disease will develop, given inheritance of a mutant version (allele) of a gene. Duchenne muscular dystrophy, for example, is a genetically transmitted disease. Virtually every male child inheriting a Duchenne mutation with his X chromosome will show characteristic early and progressive deterioration of muscle. Neurofibromatosis, defined at the level of peripheral neurofibromas and café-au-lait spots on the skin, is also a genetic disease; virtually everyone inheriting a mutant version of the NF1 gene will show these features. A few individuals inheriting a mutant NF1 gene will, however, present with a malignant cancer known as neurofibrosarcoma. Although only a small proportion, perhaps less than 10 percent, of NF1 individuals will develop a neurofibrosarcoma, this is much higher than the frequency of this type of malignancy in the non-NF1 population. Individuals with NF1 are, therefore, "predisposed" to neurofibrosarcoma; some other event or component is required before the neurofibrosarcoma can develop.

That additional component could be a characteristic of environment or life-style, a second germline-altered gene, or the stochastic occurrence of additional mutations in somatic tissue, as is common in a number of cancers. Nevertheless, inheritance of the NF1 mutation sets the stage for the ensuing event; without the NF1 mutation, the neurofibrosarcoma is very unlikely to develop. The mutant NF1 allele constitutes a genetic predisposition to neurofibrosarcoma. The crucial distinction between an inherited disorder and an inherited predisposition is that the individual with a predisposition may well never develop the disorder.

Predispositions have always been a challenge for geneticists because of their relatively low "penetrance"; only a proportion of individuals inheriting the predisposing molecular lesion actually reveal the disease end-point, thus making it difficult to see the genetic pattern. Nevertheless, it has become apparent that genetic predisposition does play an important role in a significant proportion of cases of the most common human disorders, such as cancer, heart disease, and psychiatric disease. New DNA-based technologies of human molecular genetics are making real inroads into the mysteries that have surrounded the genetics of predisposition, and we can realistically expect a flood of new knowledge describing the genetic changes that underlie specific predispositions.

Our particular problem has emerged through the study of a family we have shown to transmit a gene predisposing to colon cancer. This family now defines for us a real and pressing problem for which we must develop a solution. A brief review of the biology and genetics of colon cancer is helpful in understanding the significance of this family. The first important piece of information is that colonic polyps, a few of which will develop sporadically in many persons late in life, are believed to be the precursors of colon carcinoma. Second, in some rare families a genetic disease, familial adenomatous polyposis of the colon (FAPC), has been characterized; FAPC confers a very high risk of colon cancer because of its primary phenotype, the development of hundreds or even thousands of colonic polyps at a very early age. The likelihood of development of colon carcinoma in individuals who inherit FAPC approaches greater than 90 percent by the fifth decade of life. For this reason, most affected individuals have their colons removed before cancer can develop.

In such classic polyposis families, every individual who inherits the mutant FAPC allele will show the multiple-polyp phenotype. In these families, therefore, FAPC is a genetic disease rather than a predisposition. Recently, however, we characterized a new class of families transmitting a version of the FAPC gene that predisposes, rather than specifies, the individual carrier to an attenuated polyposis. Only a few individuals in these families will have a frank polyposis where more than 100 polyps will develop. Some families of this kind were identified as a consequence of the mapping of the classic polyposis gene to the long arm of human chromosome 5.[1,2] As a part of these earlier mapping studies and the ensuing search for the polyposis gene in the region indicated on chromosome 5, a number of very good genetic markers became available to mark the gene locus.[3]

When pedigrees were being gathered for the original polyposis mapping study, a family (K353) was identified that did not show the transmission of large numbers of polyps but did suggest the presence of an allele supporting the development of a generally quite modest, but highly heterogeneous, number of polyps. The incidence of colon cancer, however, was relatively high in this family. The clinical picture made K353 ineligible for inclusion in the FAPC-mapping study, but with DNA markers it was possible later to track the inheritance of this family's altered gene, and we showed that it was very likely the previously mapped polyposis gene.[4] The mutant version of the FAPC gene (FAPC353) transmitting in this family seems to be an attenuated, less expressive allele of the gene as regards development of polyps. Interestingly, use of the DNA markers to identify other family members who have inherited the FAPC353 allele revealed that some of these carriers have as few as one or two polyps and cannot readily be diagnosed as having polyposis. It is noteworthy that one of the family members from a previous generation who had died of colon cancer, yet had shown no polyps, was also highly likely to be a gene carrier. Taken together, the preliminary findings of this family suggest that inheritance of the FAPC353 allele confers a phenotype of a small, highly heterogeneous number of polyps and, therefore, a predisposition, at present of uncertain likelihood, to frank polyposis and/or to colon cancer. The evidence is also clear that the colon-cancer predisposition does not depend on the emergence of frank polyposis.[4]

We are now in a position to identify, with the DNA markers, a great many more individuals who have inherited the FAPC353 allele from among the several thousand living individuals in this kindred. In addition, a number of other pedigrees transmitting similar FAPC alleles have now been characterized, and inheriting individuals can be identified.

It is likely to be very important to such individuals to know that they carry an allele that strongly predisposes them to colon cancer. There is good reason to believe that frequent endoscopic screening and polypectomy for family members who carry the mutant allele would greatly reduce their chances of developing carcinoma or increase their chances of survival by early detection should a carcinoma emerge. Furthermore, it would be very important for other members of K353 and other such kindreds to know that they have not, in fact, inherited the family predisposition. Such information could free them from the morbidity and cost of the frequent endoscopic examinations recommended for at-risk family members, as well as from the pervasive fear that they are fated to die of colon cancer at an early age. We are technically able to make such a determination with high accuracy. The "catch 22" for members of K353 is that it is not feasible to inform individuals who have not inherited the FAPC353 allele of their status without revealing to others that they *have* inherited the mutant allele.

The reason that this is a catch 22 is almost purely due to the artifact of insurability. At present we believe that members of the kindred who become aware that they have actually inherited the FAPC353, colon-cancer-predisposing allele stand at considerable risk of being unable to obtain medical insurance. Under the U.S. system, access to health care is largely funded through private carriers of medical insurance, one of whose aims is to reduce overall costs by refusing to insure high-risk individuals. If specific family members were labeled as "predisposed to colon cancer," we fear that many insurance companies would come to view them as unsuitable for insurance. Therefore, we are not at present giving members of the kindred information as to individuals' genotypes, even though we may already know which among them have inherited the FAPC353 allele.

We realize, of course, that this is not a wholly new problem, but we do feel it is exacerbated by the aspect of predisposition. Many K353 members who have inherited the FAPC353 allele may never develop colon cancer, especially if they are aware of their risk and subject themselves to frequent endoscopic removal of polyps. Nevertheless, the knowledge that would be so helpful to them in dealing with their colon-cancer risk could be devastating to their lives if, as a result of the knowledge, they become uninsurable.

Colon-cancer predisposition due to the gene described here is only a beginning. It is highly likely that predispositions to other cancers, to heart disease, and to many of the other common disorders that plague us will unfold over the next decade. In many cases it may be feasible to intervene in the disease process, either through careful surveillance or through changes in life-style. We cannot turn away from the benefits to be derived from knowledge of our own individual genetic inheritance because of an artifact of our health care system. Somehow, access to health care must be protected from discrimination based on an individual's "draw" from the genetic lottery. This is not a new issue. The problem is that the number of impacted individuals is growing exponentially, and we have not yet arrived at a solution nor have we even shaped the silhouette of a solution.

A second-order problem concerns the individuals whom we identify as at-risk as a consequence of our reconstruction of pedigrees. Even without the problem of insurability, the rules for contacting such individuals remain uncertain. Should they or their physicians be notified? If they are unknowing carriers of the FAPC353 allele or a similar mutation, their risk of colon cancer may be high; the knowledge could be lifesaving. But they are not at present asking for such information; how should the situation be handled from the perspective of medical intervention? Should such people be advised of their genetic risk? The widespread screening of newborn infants for phenylketonuria constitutes a precedent suggesting that they should. This seems to be another brush-choked ethical field that has lain fallow for too long. It is time to directly confront these now very real issues.

NOTES

1. W. Bodmer, C. Bailey, J. Bodmer, et al., "Localization of the Gene for Familial Adenomatous Polyposis on Chromosome 5," *Nature* 328:614, 1987.
2. M. Leppert, M. Dobbs, P. Cambler, et al., "The Gene for Familial Polyposis Coli Maps to the Long Arm of Chromosome 5," *Science* 238:1411, 1987.
3. Y. Nakamura, M. Lathrop, M. Leppert, et al., "Localization of the Genetic Defect in Familial Adenomatous Polyposis within a Small Region of Chromosome 5," *American Journal of Human Genetics* 43:638, 1988.
4. M. Leppert, R. Burt, J. Hughes, et al., "Genetic Analysis of an Inherited Predisposition to Colon Cancer in a Family with a Variable Number of Adenomatous Polyps," *New England Journal of Medicine* 322:904, 1990.

II. Presymptomatic Genetic Diagnosis—A Worry for the United States

C. Thomas Caskey

Medical diagnostic tests are a worry for patients. It is their hope that the result be normal or, alternatively, that a therapeutic option is available in the event of an abnormal result. This is as true for the patient who undergoes a glucose tolerance test ordered because of drowsiness and polyuria as for the one who undergoes an endoscopy because of bright-red rectal bleeding. Neither patient wishes to hear the diagnosis of diabetes mellitus or colon carcinoma, but rather looks forward to a normal result. Unfortunately, disease *is* found by such diagnostic procedures, which then leads to a second level of decision—therapy. The patient found to be diabetic has a variety of therapeutic options to reduce morbidity and enhance the quality of life. The patient with early diagnosis of large-bowel carcinoma can be cured by medical intervention. We are currently accustomed to medical diagnostic tests removing uncertainty so that we can get on with our lives, or establishing a diagnosis so that we can move on to therapy options.

The new era of DNA-based diagnostic testing is changing some of that perspective on medical diagnostics. There is little doubt that DNA-based diagnostics can clarify ambiguities of clinical diagnosis such as Duchenne muscular dystrophy (DMD). Before the isolation of the DMD gene, a boy with muscle weakness would undergo a group of tests such as open muscle biopsy, electromyography, and measurement of serum enzyme levels. The combined results from these tests would indicate whether the data were consistent with the diagnosis of DMD. DNA-based methods that use a peripheral blood sample have the capacity unequivocally to identify a deletion or duplication of the DMD locus, and thus a diagnosis of the disease. The improvement in diagnostic precision for a child with weakness is welcomed by physicians and families. In addition, female carrier detection and prenatal diagnostic options are feasible using the DNA-based methods, but not by the earlier diagnostic approaches.

Nevertheless, DNA-based diagnostic methods can lead to more ambiguous situations with regard to the benefit of their implementation. It is this area that I wish to address in this chapter, which focuses on the issues involved in the application of *presymptomatic* DNA-based diagnosis of heritable disease. In order to illustrate the special new issues that have arisen and to consider solutions to these new challenges, the discussion is limited to one disorder— *adult polycystic kidney disease* (APKD). In my opinion the resolution of this challenge may provide a model for other adult-onset heritable disorders. I will introduce the discussion with a case history, based on a recent clinic referral:

> A twenty-three-year-old married man inquires about the possibility of obtaining DNA-based genotyping for APKD. His grandfather died at age forty-one with chronic renal failure. His father (age fifty) is currently undergoing chronic dialysis because of APKD. One of his brothers (age twenty-eight) has been found to have four renal cysts. The pedigree of the family is shown in Figure 10–1. Currently the consultand is asymptomatic and has not had an ultrasound study to examine for renal cysts. His major concerns regarding his undergoing testing are: how will these test results influence his purchase of life insurance and health

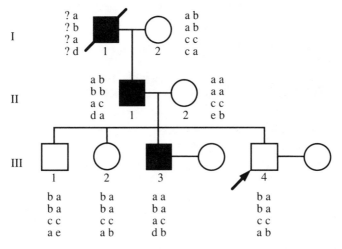

Figure 10–1. Pedigree of a family affected with APKD. Squares represent males, circles represent females, filled symbols indicate individuals with APKD. I-1, Renal failure, death at age forty-one. II-1, Age fifty, undergoing chronic kidney dialysis. III-1, Age thirty-two, without renal cysts. III-2, Age thirty, no ultrasound studies. III-3, Age twenty-eight, has four renal cysts. III-4, Consultand, age twenty-three, no ultrasound studies. Each person inherits one chromosome from each parent, and the alleles present at four loci close to the APKD locus (two on either side) are arranged vertically for each chromosome. DNA was not available from individual I-1, but since he was affected and the APKD gene is associated with the "a,b,a,d" group of alleles in this family, he was assumed to carry the "a,b,a,d" haplotype.

care coverage? He is self-employed and privately purchases both insurance policies.

Adult Polycystic Kidney Disease—Risk and Diagnosis

APKD is an autosomal dominantly inherited disorder, which occurs at a frequency of one to five per 1,000 live births and is characterized by the early onset of renal hypertension and failure.[1,2] Each child of an affected parent has a one in two chance of inheriting the disease. In the United States, an estimated 500,000 individuals are at risk and 7,000 new patients are recognized each year (NIH, information supplied by the Polycystic Kidney Research Foundation). As with many adult-onset disorders (for example, Huntington disease, myotonic dystrophy, Charcot-Marie-Tooth disease), the age at which symptoms occur varies even within a given family. The reason for such variability is not known, but to the patient this is critical because it influences the age at which medical treatment, dietary therapy, dialysis, and renal transplantation are required. At present, we lack the capacity to date precisely the age of onset of symptoms, and therefore the individual patient's decisions with regard to these various therapies. At the moment, a physician may suspect a diagnosis of APKD if more than three renal cysts are documented by ultrasonography, or if there is an established family history of the disease.

Figure 10–2 illustrates a typical course of APKD disease and incorporates information about the clinical progress of the disorder, times at which diagnosis is possible and the accuracy at those times, and the different health care policies that cover the cost of treatment. As can be seen from the figure, ultrasound detection is quite effective at diagnosis before the age at which symptoms appear,[3] but in order to have more accurate presymptomatic diagnosis before the age at which reproductive decisions are made, it is almost necessary to use the option of genetic testing.

In 1985, linkage was first detected between APKD and the 3′ hypervariable region of α-globin, localizing the APKD gene to the short arm of chromosome 16.[4] At present, there are abundant genetic markers in this region that are highly informative in terms of predicting the inheritance of the disease gene within a family.[4] There are at least two alleles at each polymorphic position close to the APKD gene; thus observation of which of these alleles are inherited in conjunction with the APKD gene in a family enables a prediction of APKD predisposition with an accuracy of greater than 95 percent (Fig. 10–1). Until the actual gene causing the disorder is cloned, it will be necessary to rely on the cooperation of family members. It is possible to perform a Bayesian calculation to estimate a probability of predisposition, based on age

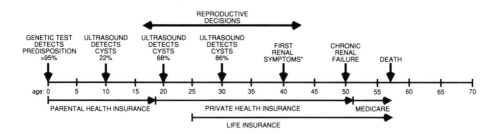

Figure 10–2. Events in the life of an APKD patient. A typical course of APKD is shown. *Hematuria, renal stones, and/or hypertension. The ability to detect renal cysts by ultrasound increases with age,[3] but genetic tests for predisposition to APKD can be performed even before birth. The APKD patient has many years of symptom-free health care and life insurance coverage before medical treatment is required. After eighteen months of chronic renal failure, the Medicare system takes over payment of medical costs from the private health carrier until the time of death.

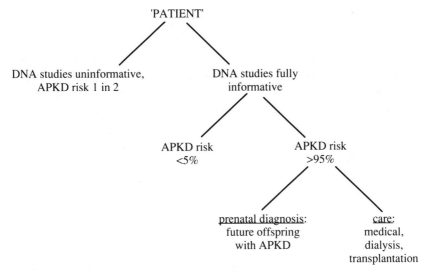

'PATIENT'

DNA studies uninformative,
APKD risk 1 in 2

DNA studies fully
informative

APKD risk
<5%

APKD risk
>95%

prenatal diagnosis:
future offspring
with APKD

care:
medical,
dialysis,
transplantation

Figure 10–3. Clarification of APKD risk. As the gene for APKD has not yet been cloned, direct detection of disease-causing mutations within the gene are not possible. Linkage studies are therefore necessary to trace the inheritance of an APKD gene within a family where at least one member is affected. If the linkage studies are uninformative, the individual with a parent affected with APKD retains a one in two risk of inheriting the disease. If the linkage studies are informative, the genetic risk can be clarified and indicate options for medical care or prenatal diagnosis.

and a genetic history. Once the gene is cloned, the possibility will exist to detect the disease-causing mutation in affected individuals, and to assay for the same mutation only in family members who wish to be tested.

Figure 10–3 illustrates that in our example family the APKD gene is inherited with the "a,b,a,d" set of alleles and that the consultand would be predicted not to develop APKD (5 percent error). The accuracy of DNA testing for APKD will undoubtedly improve rather rapidly as more is known about the genetics of the region.

If DNA is not available from enough family members, or if the genetic markers within the family are not appropriately distributed, DNA studies are said to be uninformative. In this case the risk of the patient developing APKD remains at one in two if a parent is affected. If the DNA studies are informative, the patient will know that his risk of APKD is less than 5 percent or greater than 95 percent. Thus the genetic risk has been clarified, although this does not predict the clinical course of the disease. This knowledge will permit the consultant to be better informed for discussions of family planning and prenatal diagnosis options.

Patient Care Options

Once a patient is identified with a high risk of APKD, it would be prudent to follow closely his or her renal function and blood pressure because azotemia (an excess of urea or other nitrogenous substances in the blood) and hypertension are amenable to dietary and medical therapy. If the patient is found to have a high genetic risk, routine medical surveillance should be the rule; the cost of observation is low. The care of patients with symptoms of APKD (hypertension, azotemia, renal failure) has a substantial cost, estimated in 1989 to be $694 million (direct and indirect costs) (NIH, information supplied by the Polycystic Kidney Research Foundation). The major treatment options are dietary therapy and hypertensive medications at the first signs of azotemia or hypertension, followed by kidney dialysis (six to eight years) or kidney transplantation for renal failure. Although transplantation is used less frequently than dialysis, the treatment is cheaper overall despite the cost of surgery and immunosuppression therapy. Its application is limited by supply of donor organs and the health status of the recipient.

Insurance Coverage for APKD

The substantial cost of medical care for patients with renal failure led in 1974 to the U.S. government instituting the Medicare End-Stage Renal Disease (ESRD) program, which presently costs over $400 million per year (NIH, information supplied by the Polycystic Kidney Research Foundation). Recent reduction in the funding of Medicare has displaced costs such that private carriers now bear the expense for eighteen months before Medicare funds are used. This trend is expected to continue, given the rising cost of Medicare and U.S. federal government decisions regarding limitation of Medicare expenditures. Since private health care carriers are sharing the health care cost of APKD, their attitude toward application of presymptomatic DNA testing will influence the available funding for care of patients predicted to develop the disease.

Presymptomatic DNA Testing Options

1. Do not undergo genetic testing.
2. Proceed with genetic testing for APKD.
   a. *Confidentiality.* Keep all information confidential, that is, restricted to the patient and physician.
   b. *Disclosure*
      i. Share the information, positive or negative, with the insurance corporation and Medicare.

    ii. Share the information, positive or negative, with the employer.

    iii. Share the information, positive or negative, with the family.

Consequences of Testing

1. An insurance carrier could take the position that DNA testing for pre-symptomatic disease is equivalent to screening for diabetes mellitus, hypertension, or coronary artery disease and therefore a reasonable genetic test to be conducted. The results would determine the acceptability for placement in a normal (no APKD linkage) or high (APKD linkage) insurance risk category. It should be pointed out that at present the test would be a family linkage study requiring cooperation from normal, affected, and at-risk family members. This cooperation may or may not be forthcoming.
2. An employer, either for insurance pool risk or prediction of long-term impact on job performance, could require the DNA studies in order to clarify the genetic risk.
3. A spouse may request the information, since reproduction is a joint decision.

Confidentiality

The dilemmas presented by APKD could be resolved by federal regulations. For example, it could be ruled that presymptomatic disease diagnosis should be known only to physician and patient. Would such a ruling be fair to a spouse who has admitted fear of transmitting the disease to his or her children? If one grants that a spouse has the right to the information, should that be extended to others, such as employers and insurance carriers, where employment and coverage decisions may interfere with interpretation of test results?

At present, we accept that insurance carriers have the right to determine whether we have prior existing medical conditions, including cancer, diabetes mellitus, coronary artery disease, renal failure, and so forth. We presently accept that these *prior existing conditions* influence our eligibility or rate of insurance. National health care programs such as those in Canada or the United Kingdom do not have this feature, because all citizens are provided health care coverage regardless of prior existing conditions. We need to devise a plan for the United States that takes into account private insurance carrier risk for presymptomatic disease diagnosis. I hold that presymptomatic diag-nosis is *not* disease and therefore should not be considered a prior existing condition. Thus, *an insurance carrier would not classify the risk of a pre-symptomatic APKD patient to be any different from that of the general population.* This would have the effect of freeing an individual at risk from acquiring the diagnostic information for family and personal planning. This policy is clearly to the advantage of the individual, and to the disadvantage of the insurance carrier, since the individual may purchase insurance coverage with knowledge of risk while the carrier is blind to that risk. On the other hand, it is evident from Figure 10–2 that an individual has many healthy years of insurance participation prior to the onset of health care cost, as is also true for life insurance.

Conclusions

The era of molecular medicine is upon us. We hardly blink at the use of DNA-based diagnosis for Duchenne muscular dystrophy, sickle cell disease, and cystic fibrosis, because the diseases affect children and the new methods exceed the accuracy of earlier methods. Parents generally wish their children's diseases clarified and do not involve minors in the decision. Do adults wish DNA diagnostics applied to their own presymptomatic conditions or that of their children? I have only highlighted a few of the practical issues that need resolution before we develop national policies that could impact on personal decisions.

ACKNOWLEDGMENTS

C. Thomas Caskey is a Howard Hughes Medical Institute Investigator. The assistance of Drs. Wadi Suki and Belinda Rossiter in the preparation of this manuscript is appreciated.

NOTES

1. V. E. Torres, K. E. Holley, and K. P. Offord, "General Features of Autosomal Dominant Polycystic Kidney Disease, in *Problems in Diagnosis and Management of Polycystic Kidney Disease,* edited by J. J. Grantham and K. D. Gardner, Polycystic Kidney Research Foundation, Kansas City, Mo., 1985, pp. 49–69.
2. O. Z. Dalgaard and S. Nørby, "Autosomal Dominant Polycystic Kidney Disease in the 1980's," *Clinical Genetics* 36:320–5, 1989.
3. J. C. Bear, P. McManamon, J. Morgan, R. H. Payne, H. Lewis, M. H. Gault, and

D. N. Churchill, "Age at Clinical Onset and at Ultrasonographic Detection of Adult Polycystic Kidney Disease: Data for Genetic Counselling," *American Journal of Medical Genetics* 18:45–53, 1984.

4. S. T. Reeders, M. H. Breuning, K. E. Davies, D. R. Nicholls, A. P. Jarman, D. R. Higgs, P. L. Pearson, and D. J. Weatherall, "A Highly Polymorphic DNA Marker Linked to Adult Polycystic Kidney Disease on Chromosome 16," *Nature* 317:542–4, 1985.

11

Carrier Screening for Cystic Fibrosis: A Case Study in Setting Standards of Medical Practice

Sherman Elias, George J. Annas, and Joe Leigh Simpson

Standards of medical practice have traditionally been set by physicians on the basis of their best judgment of what constitutes good patient care. This tradition may no longer be viable, both because physician values may not reflect societal values and because many other factors in addition to "good medical care" influence medical practice. Indeed, the structure of medical practice— including factors such as cost containment, the profit motive, government regulations, fear of malpractice suits, media hype, and risk-management schemes—may influence actual practice more than professional standards. This seems to be especially true in the area of genetic screening.

Physicians and their patients face a continually increasing array of diagnostic tests that demand decisions regarding clinical use. Deciding when it is the "standard of care" to offer new screening tests is perhaps the primary policy issue that the fruits of the Human Genome Project will raise. The medical profession is currently faced with another new screening test that has been heralded with banner headlines in the nation's newspapers: carrier screening for cystic fibrosis. What position should physicians take on the place of this test in their practice, and what should they tell their patients about it? This chapter summarizes the current state of the art and suggests how the nature of the test and the applicability and reliability of its results leads to conclusions that counsel its limited use in current clinical practice. The hope is that the lessons we learn in using cystic fibrosis screening as a "model" can help us develop a more generic approach to all new genetic screening tests.

Clinical Manifestations

Cystic fibrosis (CF) is the most common life-threatening autosomal recessive disease in North American Caucasians of European ancestry. In this population the disease frequency is about one in 2,500 live births and the calculated carrier frequency is about one in twenty-five individuals.[1] Common clinical manifestations of CF include meconium ileus (bowel obstruction due to excessively thick intestinal contents accumulated during fetal life), chronic obstructive pulmonary disease leading to respiratory failure, pancreatic insufficiency resulting in fatty stools and vitamin malabsorption, liver cirrhosis, and failure to thrive.[1] The clinical course of CF is highly variable, ranging from death in the neonatal period due to complications of meconium ileus to protracted survival into the fifth and sixth decades; about half the patients can be expected to live to age twenty-six or longer.[2]

Until the mid-1980s, families known to be at increased risk for CF could be offered only genetic counseling, using calculated risks based on Mendelian inheritance. In recent years, however, a number of important advances have been made. First, microvillar-intestinal-enzyme levels proved decreased in amniotic fluid of women carrying fetuses affected with CF, although the sensitivity and specificity of this assay were less than optimal.[3] Later, use of polymorphic DNA markers closely linked to the CF gene locus on chromosome 7 made prenatal diagnosis possible with a very high degree of accuracy for most couples, using either amniotic fluid cells or chorionic villi; however, DNA from the previously affected relative was necessary.[4] For close relatives of CF patients, carrier testing by DNA linkage studies could also provide a high degree of accuracy.[5]

In 1989, the CF gene was cloned and sequenced.[6] The protein presumed encoded by the CF gene has been named the cystic fibrosis transmembrane conductance regulator (CFTR) and is believed to be a membrane-bound chloride channel. However, the manner in which CFTR is involved in regulating ion conductance across the membrane of epithelial cells is still unclear.[7] Irrespective, identification and characterization of the CF gene offer strategies for increasing the accuracy of CF prenatal diagnosis beyond that possible by linkage analysis alone.

Genetics

In the United States Caucasian population of European ancestry (excluding Ashkenazic Jews and Hispanics) about 75 percent of the mutations in CF patients correspond to a three-base-pair deletion. This deletion results in the loss of a phenylalanine residue at amino acid position 508 (designated ΔF_{508}) from the 1,480-amino-acid coding region of the CF gene.* Studying 200 non-Ashkenazic Caucasian families in which a family member had CF, Lemna and colleagues[8] determined that 58.2 percent of individuals affected with CF were homozygous for the ΔF_{508} mutation, 33.7 percent were heterozygous for this mutation, and 8.0 percent were homozygous for the absence of this mutation. (Overall, the frequency of CF chromosomes was 75 percent.) The frequency of the ΔF_{508} mutation varies among different ethnic groups. For example, in Spanish and Italian populations the ΔF_{508} mutation accounts for only 49 and 43 percent of CF mutations, respectively.[9] The three-base-pair deletion appears to be much less common in the U.S. Ashkenazic Jewish population, being present in only about 30 percent of CF mutations.[10] The frequency of CF in American blacks is about one in 17,000 births[11]; however, the proportion of CF cases due to the ΔF_{508} mutation in this population is not yet well defined.

Assuming a one in 2,500 frequency of CF in the U.S. non-Ashkenazi Caucasian population, the likelihood of detecting an individual as a CF carrier can be determined, given various mutation detection rates (Table 11–1). If, for example, 85 percent of CF mutations are detectable by DNA analysis, the likelihood that an individual with a "negative" test is a carrier is reduced from an a priori risk of one in twenty-five to a new risk of one in 165. If the DNA of one parent is positive for a mutation but the DNA of the other parent negative, the risk for the couple having a child with CF is (1, or known heterozygote)(1/165)(1/4), or 1/660. If DNA of neither parent tests positive for a mutation, the risk of having a child with CF is (1/165)(1/165)(1/4), or about 1/109,200. Contemplating population-based screening necessitates deriving the likelihood of detecting couples who are actually CF carriers, which can be calculated as a function of the mutation detection rate (Table 11–2). At an 85 percent mutation detection rate, for example, both partners will be identified as carriers in 72.3 percent of the cases; in 25.5 percent of couples only one of the partners will be identified as a CF carrier; in 2.2 percent neither partner will be identified as a CF carrier.[12]

---

*In the context of this chapter, the term "mutation analysis" refers to abnormalities of the CF gene itself, demonstrable by DNA analysis (for example, ΔF_{508}). By contrast, the term "linkage analysis" refers to restriction fragment length polymorphism (RFLP) DNA markers that are linked to the CF gene but not within the actual gene. Thus, RFLP analysis only indirectly provides information concerning the presence or absence of the CF mutation.

Table 11-1. Cystic Fibrosis-Related Risks after Mutation Analysis for Carrier Status

Percent of cystic fibrosis mutations detectable	Carrier risk for person with negative test	Risk of cystic fibrosis in offspring for couples tested	
		One parent positive*	Neither parent positive
0	1 in 25.5	NA	1 in 2,500
70	1 in 82.7	1 in 331	1 in 27,400
75	1 in 99	1 in 396	1 in 39,200
80	1 in 124	1 in 494	1 in 61,000
85	1 in 165	1 in 661	1 in 109,200
90	1 in 246	1 in 984	1 in 242,100
95	1 in 491	1 in 1964	1 in 964,400

*NA denotes not applicable.

Reproduced with permission from Lemna et al.[8]

Table 11-2. Potential for the Detection of Couples at Risk with the Use of Population-Based Screening for Carrier

Percent of cystic fibrosis mutations detectable	Detection of carrier status		
	Both parents (%)	One parent (%)	Neither parent (%)
70	49.0	42.0	9.0
75	56.3	37.5	6.2
80	64.0	32.0	4.0
85	72.3	25.5	2.2
90	81.0	18.0	1.0
95	90.3	9.5	0.2

Reproduced with permission from Lemna et al.[8]

A number of clinical scenarios can be envisioned for the U.S. non-Ashkenazi Caucasian population, assuming the current CF mutation detection rate of 85 percent.

Previous Child Affected with CF

1. *Both parents heterozygous for a detectable CF mutation, with or without affected child available.*

Barring the rare occurrence of a new mutation, couples who have a child with CF have a one in four recurrence risk in each subsequent pregnancy. If the parents are heterozygous for a detectable CF mutation, prenatal diagnosis by direct mutation analysis using either chorionic villi or amniotic fluid cells is the method of choice (in terms of speed and reliability). In couples who have had an affected child this can be expected in 0.85 × 0.85, or 72.3 percent of cases. When the CF mutation in each parent can be detected, there is almost 100 percent accuracy for identifying an affected fetus. In such cases DNA from a previously affected child is not required.

2. *One or both parents carrying CF mutations that are not yet detectable; DNA from affected child available.*

The affected child could be a compound heterozygote; that is, one parent could contribute a chromosome 7 with a detectable mutation such as ΔF_{508} whereas the other parent could contribute a chromosome 7 with a (different) mutation in the CF gene that cannot yet be directly detected. Alternatively, a child could be affected with CF as result of each parent contributing a chromosome 7 with a mutation in the CF gene neither of which can yet be detected. In both these situations, prenatal diagnosis can still be offered with a high degree of accuracy, provided DNA from the affected child is available. A combination of ΔF_{508} mutation analysis and RFLP linkage analysis is used in the former circumstance, whereas linkage analysis alone is used in the latter.

3. *Parents with discrepant mutations in the CF gene; DNA from affected child not available.*

If one parent carries a detectable CF mutation and the other parent does not, DNA from the fetus can be analyzed to determine whether it inherited the mutation from the parent with the detectable CF mutation. If not, the fetus is unaffected. For example, if a ΔF_{508} carrier parent *did* transmit the CF gene to the fetus, there is a one in two risk that the fetus is affected. Under these circumstances amniotic fluid microvillar-intestinal-enzyme analysis would be particularly useful in predicting fetal CF status. Enzyme analysis carries an estimated false positive rate of 2 percent and false negative rate of 8 percent.[13]

Close Relative with CF

An unaffected sib of an individual with CF has a two in three likelihood of being a CF heterozygote and a one in three likelihood of being homozygous normal. An uncle or aunt of an individual with CF has a one in two likelihood of being a CF heterozygote and a one in two likelihood of being homozygous normal. A first cousin to an individual with CF has a one in four likelihood of being a CF heterozygote and a three in four likelihood of being homozygous normal. Thus, if an individual has a close relative with CF (sib, niece, uncle, or first cousin), that individual should be offered carrier testing using mutation analysis or linkage analysis. If the parent having a close relative with CF is indeed found to be a carrier, mutation analysis of the spouse is indicated. If both parents are determined to be carriers, their risk for having offspring with CF is one in four; prenatal diagnosis would be appropriate. If the one parent carries the detectable CF mutation (presumably the high-risk parent) and the other parent does not carry a detectable mutation, the couple's risk for having a child with CF is one in 661. If the carrier parent *did* transmit the CF gene to the fetus, there is still only a one in 335 risk that the fetus is affected. Under such circumstances prenatal diagnosis could not at present provide an unequivocal determination of fetal status. Even if microvillar-intestinal-enzyme analysis was employed, Bayesian calculations based on known false positive (2 percent) and false negative (8 percent) rates reveal that an abnormal result would yield only a 19 percent probability that the fetus is affected.[14] Conversely, there is an 81 percent probability that microvillar-intestinal-enzyme analysis would represent a false positive result.

Population-Based Cystic Fibrosis Carrier Screening

As discussed above, in the U.S. non-Ashkenazic Caucasian population, about 75 percent of heterozygotes will have the ΔF_{508} mutation. Therefore, as shown in Table 11–2, only about 56 percent of those couples in which both partners are CF carriers would be identified by screening for the ΔF_{508} mutation alone. With additional CF mutations being discovered, the percentage of couples that can be identified in which both partners are CF carriers is increasing. If, for example, 95 percent of carriers could be detected, 90 percent of couples at risk would be identified (Table 11–2). It was initially hoped that the remaining total number of CF mutations would be relatively small, thus allowing most, if not all, CF carriers to be identified efficiently. Unfortunately, this has not proved to be the case; numerous additional mutations have already been identified, each individually rare. As alluded to earlier, the current CF mutation detection rate is approximately 85 percent. If population-based CF screening were undertaken now, in about one in twenty couples one partner would be identified to be a CF carrier and the other partner would not carry a detectable mutation. Failure of the other partner to have a detectable mutation does *not,* however, exclude that partner being a CF carrier. Rather, the likelihood is merely reduced from one in twenty-five to one in 165 (Table 11–1). In such couples the risk for having a child with CF is about one in 661, but prenatal diagnosis cannot presently provide an unequivocal determination of fetal CF status even with the combination of amniotic fluid microvillar-intestinal-enzyme testing and RFLP linkage analysis.

Public Policy for Carrier Screening

In March 1990, the National Institutes of Health (NIH) Workshop on Population Screening for the Cystic Fibrosis Gene[15] was convened to make suggestions regarding the introduction of cystic fibrosis screening into medical practice. The group concluded that "pilot programs" investigating research questions in the delivery of population screening for CF are "urgently needed." The NIH panel listed the following five guidelines, that are already widely accepted as applicable to any widespread population screening. We believe each merits elaboration in the context of CF screening:

1. *Screening should be voluntary and confidentiality must be assured.*

   The National Academy of Sciences recommended in 1975 that "participation in a genetic screening program should not be mandatory by law, but should be left to the discretion of the person to be tested or, if a minor, of the parents or legal guardian."[16] The President's Commission on Bioethics strongly endorsed this recommendation, noting that "some individuals are likely to choose not to be screened for CF, and their ability to make that choice must be safeguarded."[17]

   Confidentiality enhances the trust implicit in the doctor-patient relationship, and medical information should not be disclosed to third parties without informed consent. As stated by the President's Commission: "Because of the potential for misuse as well as unintended social or economic injury, information from genetic testing should be given to people such as insurers or employers only with the explicit consent of the person screened."[18] We can foresee no circumstances under which confidentiality should be breached in CF carrier screening programs. Even if an individual identified as a carrier refuses to allow this information to be conveyed to a sibling, the a priori risk of that sibling having a child with CF is less than 1 percent.

2. *Screening requires informed consent;*
3. *Providers of screening services have an obligation to assure that adequate education and counseling are included in the program.*

   As important as autonomy is, it is illusory if the individual does not know to what he or she is being asked to consent. Accordingly, there is need to develop prescreening educational material that adequately explains, in lay terms, the benefits of screening as well as the limitations and risks of screening, such as loss of insurability and adverse psychosocial effects. This information must, of course, be made available to the individual *before* consent for screening is sought. In addition, professional counseling regarding the significance of positive screening results and reproductive options must be available when results are made known to the individual.

The goal of genetic screening and counseling is "to make people into informed decision makers about their genetic constitution, to the extent it is relevant to choices about their own well-being or that of their family."[19] The success of population-based CF carrier screening will depend on the practicing medical community becoming knowledgeable and supportive of the entire process as well as on educating and involving the public. Before population carrier screening begins, informational campaigns at national and local levels must be in place to educate the public concerning the objectives and availability of screening. It has been estimated, for example, that if screening were done on three million couples a year who are contemplating having children, and if ten minutes were spent counseling them before the test and an hour spent counseling at-risk couples after the test, 651,000 hours of counseling would be required to meet the screening-generated demand.[20] This would require more than 17 weeks per year for each of the approximately 1,000 certified genetic counselors and clinical geneticists, just for CF screening of this limited population.[21] The alternative is that primary providers (for example, obstetrician–gynecologists and family practitioners) possess the requisite knowledge and counseling skills.

As the number of possible genetic screening tests proliferates, it will be necessary to devise new strategies to educate and help individuals to make the decision to utilize specific screening tests. These may ultimately involve such interventions as interactive videotapes that can be played at home and in physician waiting rooms, as well as the use of specially trained nonphysician counselors in clinics and schools.

The NIH panel concluded that the current difficulty with CF screening (that is, only 75 percent of carriers are detectable) "would be substantially reduced if testing could detect at least 90 to 95% of carriers." Although there is certainly a degree of arbitrariness about any cutoff point for inclusive detection, this degree of inclusiveness (90 to 95 percent) seems reasonable if one of the major benefits of screening is excluding individuals from the at-risk category. First, without a screening process that meets this degree of inclusiveness, the amount of reassurance one can provide a couple is substantially reduced. Second, the result may be to foster both fear and stigmatization. Third, at a detection rate of 75 percent, the costs of population screening are relatively high compared to the benefits. Wilfond and Fost estimate that at this rate 11,100 couples must be screened to detect ten at-risk couples. Assuming 80 percent of these families choose prenatal diagnosis, two (25 percent) will be expected to have an affected fetus.[22] If half of these (one) terminate the pregnancy, one of two CF births will be avoided. At $200 for screening and counseling per couple, the cost of avoiding one CF birth under this scenario is $2.2 million. Even if the detection rate were increased to 95 percent, and the cost decreased to $70 per couple, the cost would still be $1 million per CF birth avoided, and this does not take into account costs incurred by effects arising from error or stigmatization.[23]

On the other hand, individual members of the public may feel that a lower mutation detection rate is acceptable for individual testing (versus population screening), thus rejecting the NIH panel's perhaps paternalistic conclusion. A flexible approach thus seems prudent and reasonable. If a highly motivated individual or couple without a family history requests CF carrier testing, the current state of the art (as set forth above) should be carefully explained by the physician, either directly or through a genetic counseling center. If, after this explanation, the individual still wants to be tested, it is not unreasonable for the physician to comply with this request.

4. *Quality control of all aspects of the laboratory testing including systematic proficiency testing is required and should be implemented as soon as possible.*

Accurate screening and counseling will require reliable test results. It is thus important that national standards for laboratory testing be established, that the laboratories involved in performing CF testing be monitored in some reasonable manner, and that the results of such monitoring be made available to the medical profession and the public.[24]

5. *There should be equal access to testing.*

We agree with the NIH panel and others that routine screening of the general population is premature at this point. If, however, successful pilot programs demonstrate that the benefits outweigh the costs, and that counseling can be done effectively, the question will be, who should be screened? The NIH panel stated that "the most appropriate group for population screening is reproductive aged individuals," with primary health providers being "the optimal setting." Whether CF screening should be covered by public insurance is a generic issue that society is currently wrestling with: of what should the "minimum benefit package" consist? Unless the cost of the test can be reduced from its present $150 to $200, it may be difficult to justify population screening covered by private insurance, even more difficult to justify population screening covered by public insurance, and even more difficult yet to justify screening American blacks. Additional cost and benefit data will be needed to make this assessment.

Ideally CF carrier screening would be performed prior to pregnancy, at a time when couples still have such options as artificial insemination by donor, adoption, or not having children at all. Some couples might choose to be screened, but even when identified as CF carriers might elect to take their chances with the knowledge that there is "only" a 25 percent risk of having an affected child. However, screening adults for heterozygosity is generally useful only if prenatal diagnosis is available. At present it can be assumed that couples who elect to be screened in which both partners are identified to be heterozygotes for CF would usually proceed to prenatal diagnosis with the intent of pregnancy termination if the fetus is found to be affected.

Inevitably, this may be a difficult choice for some couples. This is not only because many affected children cope with their disease admirably, even living four decades, but also because the CF gene having been cloned and its gene product (CFTR) predicted has brought new optimism that either somatic gene therapy or medical therapy targeted toward the underlying abnormalities will provide a "cure" or significant amelioration for patients with CF in the foreseeable future.

Prior to amniocentesis or chorionic villus sampling (CVS), couples who are CF carriers require knowledgeable counseling by experts in CF who can accurately detail in unbiased fashion the likely clinical course of the disease, as well as the medical progress being made in its treatment. It is the hope that eventually CF will no longer be an indication to consider pregnancy termination. At present, however, the final decision regarding pregnancy termination must rest with the accurately counseled couple.

Setting Standards of Care for Genetic Screening

At the height of the second "medical malpractice insurance crises" in the mid-1980s, it appeared that physicians might be ready to set standards based on their fear of litigation, rather than on their view of patient benefit. The most notorious example was in the field of prenatal screening and involved the use of maternal serum alpha-fetoprotein screening (MSAFP); it provides a useful overview of the legal issues relevant to standard setting.[25]

Neural tube defects (NTDs) are genetic disorders that occur with an overall incidence of one to two every 1,000 live births in the United States. The two major types are anencephaly and spina bifida. Couples who have had a child with one of these conditions have a 2 to 3 percent risk of recurrence and account for 5 to 10 percent of all NTDs. Unfortunately, the remaining 90 to

95 percent of NTDs occur in families without any prior medical history. To be effective, a screening test for NTDs must therefore be used routinely on all pregnant women.

Alpha-fetoprotein is a major fetal serum protein secreted by the fetal kidneys and is normally present in amniotic fluid in measurable amounts. It may also, however, enter the amniotic fluid directly from exposed membrane surfaces on the fetus, as in anencephaly or open spina bifida. There is also an association between elevated levels of MSAFP and NTDs; measuring second-trimester concentration of MSAFP has been shown to be an effective means of identifying pregnant women who are at risk of having a fetus with an NTD.

Following counseling and obtaining of informed consent, a blood sample is taken from the pregnant woman some time between fifteen and twenty weeks of gestation as calculated from the first day of the last menses. In patients showing an elevated MSAFP (approximately 5 percent of those screened), the test may be repeated. If the second MSAFP level is again elevated, an ultrasonographic evaluation is advised to detect conditions other than NTDs, such as erroneous estimation of gestational age, multiple pregnancies, fetal death or threatened abortion, Rh disease, and certain rare congenital abnormalities that are also associated with elevated MSAFP.

In less than half of this group (1 to 2 percent of the original number screened) no explanation for the MSAFP elevation is found, and an amniocentesis for amniotic fluid AFP and acetylcholinesterase is recommended. Approximately 5 to 10 percent of those undergoing amniocentesis show elevated AFP levels, most associated with NTDs. Properly administered, this series of tests detects 80 to 90 percent of all fetuses with anencephaly and 70 to 90 percent of all fetuses with open spina bifida. More recently, low levels of MSAFP, especially combined with low maternal serum unconjugated estriol and high human chorionic gonadotropin, have been capable of identifying up to 60% of pregnancies involving fetal Down syndrome in women under age thirty-five.[26]

Reasons for not rushing into routine MSAFP screening are evident from this description. High-quality laboratory, counseling, ultrasound, and amniocentesis services must be available and accessible across the country. In the absence of such services, MSAFP could simply increase cost and parental anxiety (since the vast majority of elevated MSAFP sample will turn out not to be associated with NTDs or any other abnormality) and possibly lead to unnecessary abortions.

These and other reasons led the American College of Obstetricians and Gynecologists (ACOG) to oppose FDA approval of AFP test kits, and to tell its members in October 1982 that although the test was useful for those patients who had had a previously affected child, "routine maternal serum AFP screening of all gravida is of uncertain value":

> Maternal serum AFP screening should be implemented only when it can be performed *within a coordinated system of care that contains all the requisite resources* and facilities to provide a safeguard essential for ensuring prompt, accurate diagnoses and appropriate follow-through services. When such coordination of resources and services is not possible, the risks and costs appear to outweigh the advantages and the program should not be implemented.[27]

Nonetheless, with no stated justification other than the 1983 FDA approval of commercial kits for the radioimmunoassay of AFP, ACOG members received a May 1985 "Alert" from the College's Department of Professional Liability entitled "Professional Liability Implications of AFP Tests." The Alert stated in part:

> It is now imperative that you investigate the availability of these tests in your area and familiarize yourself with the procedure, location and mechanism of the follow-up test to screen for neural tube defects. It is equally *imperative that every prenatal patient be advised of the availability of this test* and that your discussion about the test and the patient's decision with respect to the test be documented in the patient's chart.[28]

The rationale for this advice was not medical, but legal: to give the physician "the best possible defense" in a medical malpractice suit premised on the birth of a baby with an NTD. ACOG's Department of Professional Liability had been very agitated about what it saw as a crisis of medical malpractice insurance unavailability and unaffordability, a crisis that has reportedly led about 10 percent of ACOG's Fellows to give up obstetrics altogether and almost 20 percent to drop high-risk obstetrics. But if malpractice litigation is a problem, letting lawyers set medical standards of care is hardly the solution.

The general rule has always been that "good medicine is good law." Physicians who follow good and accepted medical standards in their practice will be found in compliance with the general standard of "reasonable prudence" for physicians and thus cannot be found negligent in a malpractice suit. It may also be said of the ACOG Alert that "bad law is bad medicine." ACOG's scientific committees, not its legal liability department, should so advise its members on good medical practice. Indeed, this policy appears to have been scrupulously followed since 1985.

Perhaps ACOG's Lawyers were in 1985 thinking about the case of *Helling v. Carey.*[29] Two ophthalmologists were found negligent for not having performed a glaucoma test on a young woman whose glaucoma, when it was finally discovered, had progressed to a point where her peripheral vision was lost and her central vision severely reduced. In their defense, the physicians argued that it was the standard of care in the ophthalmology community not to routinely screen patients under the age of forty for glaucoma because the incidence of the disease in this population (about one in 25,000) was low. On appeal from a verdict for the physicians, the Supreme Court of Washington noted that the standard of care is judged on the basis of "reasonable prudence." Even though ophthalmologists did not routinely offer this test to patients under forty years of age, the court decided that not to so offer it was a failure of reasonable prudence. The glaucoma test, the Court found, was simple, inexpensive, safe, accurate, and could detect an arrestable and "grave and devastating" disease.

The Court quoted with approval the past statements of Justices Oliver Wendell Holmes and Learned Hand, both to the effect that while reasonable prudence is usually measured by custom or what is in fact done, custom can never be its sole measure. As Justice Hand put it in the *T. J. Hooper* case:

> [A] whole calling may have unduly lagged in the adoption of new and available devices. It never may set its own tests, however persuasive be its usages. *Courts must in the end say what is required: there are precautions so imperative that even their universal disregard will not excuse their omission.*[30]

But what would this same court have said about MSAFP screening? True, the initial screening is relatively inexpensive and safe. Yet accuracy requires highly trained professionals to perform a complex series of follow-up tests, including ultrasound and amniocentesis. And even with this, there will be false negatives. Nor can this series of tests be fairly characterized as "simple." It does detect a serious condition, one that can even be described as "grave and devastating," but the condition cannot be "arrested" or "treated" except by aborting the affected fetus.

The lessons for CF screening should be obvious: medical organizations should set standards based on medical benefit, not fear of lawsuits. Setting standards on the latter basis causes more harm than good, to both providers and physicians. Indeed, ultimately ACOG's scientific committees did recommend that MSAFP screening be routinely offered to all pregnant women.

The Obligations of Obstetrician–Gynecologists for CF Screening

Screening offered as part of more generalized health care services (for example, through high-school, college, or family-planning clinics) would encourage preconceptual CF carrier screening. However, the reality is that for most couples the first opportunity for carrier screening will occur only after a pregnancy has occurred. Thus, as with other types of genetic screening (for example, maternal serum alpha-fetoprotein, advanced maternal age, hemoglobinopathies, Tay-Sachs disease), the primary responsibility for providing CF carrier screening will thus rest with obstetrician–gynecologists.

One important recommendation of the NIH panel was that CF screening "should be offered to all individuals and/or couples with a family history of cystic fibrosis." On the other hand, for many of the reasons already detailed, the group recommended against screening those individuals with a negative family history "under the current circumstances." This recommendation is consistent with recent professional organization statements. The ACOG Committee on Obstetrics: Maternal and Fetal Medicine reviewed and agreed with a 1990 policy statement written by the American Society of Human Genetics, which states the following:

> First, carrier testing should be offered to couples in which either partner has a close relative affected with CF. Second, one or a few federal, foundation, or privately supported pilot programs should be conducted as soon as possible in order to gather more data regarding laboratory, educational and counseling aspects of screening. Third, there is an immediate need for centralized quality control of laboratories conducting these tests. Fourth, it will be appropriate to begin large-scale population screening in the foreseeable future, once the test detects a larger proportion of CF carriers and more information is available regarding the issues surrounding the screening process. Until that time it is considered premature to undertake population screening.
>
> Finally, it is the position of the American Society of Human Genetics that *routine* CF carrier testing of pregnant women and other individuals is NOT yet the standard of care in medical practice.[31]

Current "standard of care" for obstetrician–gynecologists includes taking a genetic history and offering (not requiring) prenatal diagnostic testing to couples who have had a previous child with CF.[32] The question is how we decide whether CF screening should be routinely recommended for individuals of reproductive age who have a negative family history. The President's Commission on Bioethics,[33] for example, recommended that "public screening programs should not be implemented until they have first demonstrated their value in well-conducted pilot studies." The NIH workshop panel concurred with this recommendation, stating:

> Pilot programs investigating research questions in the delivery of population screening for CF carriers are urgently needed. These programs should address clearly defined questions including effectiveness of educational materials, level

of utilization, laboratory aspects, counseling issues, costs, and beneficial and deleterious effects of screening.[34]

We agree. Good science and solid educational and counseling strategies must be in place before CF screening is routinely offered to those with a negative family history. CF carrier screening should be deemed "standard of care" only after the appropriate scientific committees (for example, the Committee on Obstetrics: Maternal and Fetal Medicine of ACOG) have reviewed all relevant scientific data and educational strategies and concluded that offering such screening is likely to provide more benefit than harm to individuals and couples. This will help assure that CF carrier screening will be introduced in a coordinated and effective manner.

Under the influence of aggressive private companies marketing CF tests, individual practitioners may be tempted to offer routine CF screening. The medical profession can still be effective in managing the introduction of CF testing and screening into clinical practice. This is important because ad hoc approaches to genetics screening, such as those utilized with sickle cell anemia, Tay-Sachs disease, MSAFP, and now CF screening are ultimately unsatisfactory. As the Human Genome Project continues, tens, if not hundreds, of new genetic screening tests may compete for introduction into "routine" clinical practice. It is past time to begin to develop guidelines for determining how we should judge each such new test, which tests should be routinely offered, and who should pay for them. Current procedures in which professional associations like ACOG and expert panels at NIH make recommendations are cumbersome and untimely and fail to contemplate or allow effective participation by the public. Given that the public will ultimately be asked to accept or refuse such screening, and given that the ultimate objective of all such screening is to benefit the public, effective steps should be taken to involve the public in deliberations concerning which screening tests should routinely be offered.

NOTES

1. T. F. Boat, M. J. Welsh, and A. L. Beaudet, "Cystic Fibrosis," in *The Metabolic Basis of Inherited Disease,* 6th ed., edited by C. R. Scriver, A. L. Beaudet, W. S. Sly, and D. Valle, McGraw-Hill Book Co., New York, 1989, pp. 2649–80.
2. Ibid.
3. D. J. Brock, H. A. Clarke, and L. Barron, "Prenatal Diagnosis of Cystic Fibrosis by Microvillar Enzyme Assay on a Sequence of 258 Pregnancies," *Human Genetics* 78:271, 1988.
4. L. C. Tsui, M. Guchwald, D. Barker, et al., "Cystic Fibrosis Locus Defined by a

Genetically Linked Polymorphic DNA Marker," *Science* 230:1054, 1985; and R. White, S. Woodward, M. Leppert, et al., "A Closely Linked Genetic Marker for Cystic Fibrosis," *Nature* 318:382, 1985.

5. A. L. Beaudet, G. L. Feldman, S. D. Fernbach, G. J. Buffoone, and W. E. O'Brien, "Linkage Disequilibrium, Cystic Fibrosis, and Genetic Counseling," *American Journal of Human Genetics* 44:319, 1989.

6. J. M. Rommens, M. C. Lannuzzi, B. Kerem, et al., "Identification of the Cystic Fibrosis Gene: Chromosome Walking and Jumping," *Science* 245:1059, 1989; J. R. Riordan, J. M. Rommens, B. Kerem, et al., "Identification of the Cystic Fibrosis Gene: Cloning and Characterization of Complementary DNA," *Science* 245:1066, 1989; and B. Kerem, J. M. Rommens, J. A. Buchanan, et al., "Identification of the Cystic Fibrosis Gene: Genetic Analysis," *Science* 245:1073, 1989.

7. Riordan et al., 1989.

8. W. K. Lemna, G. L. Feldman, B. Kerem, et al., "Mutation Analysis for Heterozygote Detection and the Prenatal Diagnosis of Cystic Fibrosis," *New England Journal of Medicine* 322:291, 1990.

9. X. Estivill, M. Chillon, T. Casals, et al., "ΔF508 Gene Deletion in Cystic Fibrosis in Southern Europe," *Lancet* 2:1404, 1989.

10. Lemna et al., 1990.

11. I. L. Kulczycki and V. Schauf, "Cystic Fibrosis in Blacks in Washington, D.C.: Incidence and Characteristics," *American Journal of Diseases of Childhood* 127:64, 1974.

12. Lemna et al., 1990.

13. Ibid.

14. Ibid.

15. "Statement from the National Institutes of Health Workshop on Population Screening for the Cystic Fibrosis Gene," *New England Journal of Medicine* 323:70, 1990.

16. National Academy of Sciences, *Genetic Screening: Programs, Principles and Research,* National Academy of Sciences, Washington, DC, 1975.

17. President's Commission for the Study of Ethical Problems in Medicine and Biomedical and Behavioral Research, *Screening and Counseling for Genetic Conditions,* United States Government Printing Office, Washington, D.C., 1983.

18. Ibid.

19. NIH Workshop, 1990.

20. B. S. Wilfond and N. Fost, "The Cystic Fibrosis Gene: Medical and Social Implications for Heterozygote Detection," *Journal of the American Medical Association* 263:2777, 1990.

21. Ibid.

22. Ibid.

23. Ibid.

24. N. A. Holtzman, *Proceed with Caution,* Johns Hopkins University Press, Baltimore, 1989.

25. Much of the discussion of MSAFP screening is from S. Elias and G. J. Annas, *Reproductive Genetics and the Law,* Yearbook Medical Publishers, Chicago, 1987.

26. N. J Wald, H. S. Cuckle, J. W. Densem, et al., "Maternal Serum Screening for Down's Syndrome in Early Pregnancy," *British Medical Journal* 297:883, 1988.

27. ACOG, Technical Bulletin No. 67, "Prenatal Detection of Neural Tube Defects," ACOG, Washington, D.C., 1982 (emphasis added).
28. Reprinted in Elias and Annas, 1987, p. 74 (emphasis added).
29. *Helling* v. *Carey,* 83 Wash. 2d 514, 519 P. 2d 981 (1974).
30. Ibid. (emphasis added by the court).
31. "The American Society of Human Genetics Statement on Cystic Fibrosis Screening," *American Journal of Human Genetics* 46:393, 1990.
32. American Academy of Pediatrics and American College of Obstetricians and Gynecologists, *Guidelines for Perinatal Care,* 2nd ed., American College of Obstetricians and Gynecologists, Washington, DC, 1988.
33. President's Commission, 1983.
34. NIH Workshop, 1990.

12

What We Still Don't Know about Genetic Screening and Counseling

James Sorenson

In 1972 Drs. Fredrick Hecht and Lewis Holmes published a letter in the *New England Journal of Medicine* titled "What We Don't Know about Genetic Counseling."[1] Their comments accompanied an article by Leonard and colleagues titled "Genetic Counseling: A Consumer's View," which reported data on how clients respond to genetic counseling.[2] Hecht and Holmes provided a list of questions about genetic counseling that covered a broad spectrum of issues, ranging from the types of training necessary for effective counseling, to questions about the best methods for counseling, to questions about how clients use the information provided in counseling. It has been almost twenty years since Hecht and Holmes raised their questions. In the intervening period many articles and several books have been published on genetic counseling as well as genetic screening, addressing some of their questions.[3] This research has begun to provide answers to some of the questions, but not all.

Initiation of the Human Genome Project, with its anticipated impact on applied human genetics in general and genetic screening and counseling in particular, has renewed interest in questions of the kind Hecht and Holmes raised and in answers to such questions. While it is not possible to foresee all the ways in which developments in the Human Genome Project may impact on applied human genetics, it is highly probable that they will increase the scope and magnitude of (a) newborn genetic disease screening; (b) heterozygote screening; (c) prenatal diagnosis; and (d) presymptomatic genetic disease screening. It also appears probable that developments will lead to a new form of genetic screening, susceptibility screening, in which a person will be made aware of being "at risk" for developing such diseases as cancer or cardiovascular disease because of his or her genetic endowment or environmental exposures. Genetic counseling is or will be central to the delivery of all these forms of screening. Given this, as the Human Genome Project begins to bear fruit, questions about the impact and effectiveness of genetic counseling will take on increased interest.[4]

Some Major Questions

The questions asked by Hecht and Holmes covered many topics. In contrast, the article by Leonard and colleagues focused almost exclusively on the topic of client responses to counseling, or, as Leonard et al. labeled it, a "consumer's view." An examination of the published literature on genetic counseling and screening over the past twenty years suggests that while some attention has been given to issues such as counselor training and access to counseling and screening services, most attention has been devoted to studying clients receiving screening and/or counseling, that is, at least in part a "consumer's view." In reviewing this literature, it is possible to organize the many questions raised about counseling and its impact into five broad areas:

1. How effective is genetic counseling in providing clients with genetic and medical information to enable more informed reproductive decisions?
2. How useful do genetic counseling clients find the information provided in making reproductive decisions and taking actions subsequent to counseling?
3. Does genetic counseling impact on a client's psychological status?
4. Does genetic counseling impact on the reproductive plans and behavior of clients?
5. Does genetic counseling impact on clients' personal relationships, marital satisfaction, and family stability?

These are, of course, very complicated questions. For some of these, research has begun to provide tentative answers, but for others, almost no research has been conducted.

A review of these five questions reveals an inherent ordering or model of how one might expect genetic counseling to have an effect. More specifically, if counseling is to have an impact, (1) clients must be provided with and learn some genetic and medical information; (2) if the information is to have an impact on clients, they must see its personal relevance for them; (3) once learned and personally understood, such information can impact positively or negatively on how clients feel about themselves and their situation; such psychological responses can mediate between the factual knowledge clients have and their behavior; (4) what clients have learned and how they interpreted it for themselves could shape reproductive behavior, which in turn (5) could impact on relationships and their stability. This is a highly oversimplified view, of course, and does not do justice in its unrelenting linear logic to the recursive and complex reality of human problem solving. It does, however, provide a scheme for reviewing studies on genetic screening and counseling. Perhaps more important, the five questions listed above parallel closely major elements of genetic counseling identified in the 1975 American Society of Human Genetics definition as key objectives of counseling.[5]

In the remainder of this chapter I will discuss each of these questions. I will comment on conceptual and research problems in studying such questions, very briefly comment on what research suggests is presently known in response to these questions, and make some comments and suggestions regarding possible future studies.

Education in Counseling

The 1975 definition of genetic counseling makes it clear that the education of clients in basic medical and genetic facts is a cornerstone of counseling. Without much specificity, the definition identifies a limited set of medical (diagnosis, treatment) and genetic (mode of inheritance, recurrence risk) "facts" that clients should be given. If one draws on the decision analytic literature and views a primary issue confronting most couples in counseling as that of making a reproductive decision under conditions of risk, then at a minimum two sets of medical and genetic information are required—risk information and burden information. In the research literature on genetic counseling, considerable attention has been given to assessing client risk knowledge and risk interpretation. More limited attention has been given to assessing client conceptions of the burdens associated with genetic problems. Moreover, there has been almost no research assessing how effective counseling is at improving client understanding of disease burden. Some attention has been given to how well clients learn the diagnosis, but again, very little attention has been paid to clients' understanding of treatment information.

Concepts such as treatment and disease burden are enormously complex, of course, and not only difficult to conceptualize, but perhaps even more difficult to measure. This, no doubt, has contributed to their highly limited role in existing research. Nevertheless, they are important concepts needing development. Client conceptualizations of disease burden can influence not only their decision to risk a pregnancy when prenatal diagnosis is not available, but also their interest in using prenatal diagnosis when it is an option. If the Human Genome Project leads to greater diagnostic and screening possibilities, as well as improved treatment, then client perceptions of disease burden may play a significant role in shaping utilization rates of new genetic technologies developing out of the Project.

A large number of studies have reported data on client learning in counseling. This literature has been extensively reviewed recently.[6] The impression one gains from this literature is that, in terms of client learning there is wide variability in reported rates of learning. In general, however, genetic counseling appears as successful as many other medical encounters in educating clients.[7] More precisely, with respect to such information as risk estimates and diagnosis, substantial majorities of clients in the more carefully controlled studies can correctly report their risks and diagnosis soon after, as well as several months after, counseling. However, it is also the case that sizable minorities cannot, meaning that there is significant room for improvement in education in counseling.

With few exceptions, most of the existing research consists of observational studies. This means that the researchers record levels of client knowledge before and after counseling and impute changes to counseling. In a few cases, such changes have been compared to "control" groups, and in even fewer cases, researchers have assessed different methods of education, such as counselor-based, client-centered, and video plus counselor education.

The present level of research on the effectiveness of genetic counseling as an educational encounter indicates that it is sometimes successful, sometimes not. Given the variability in genetic counseling, from clinic to clinic and often from practitioner to practitioner, this is not surprising. What is not clear at this time is what works well and what works less well in counseling. Additional research is needed to clarify this issue. In contrast to observational studies, what is needed are "intervention" studies. In such investigations researchers develop theoretically and empirically informed interventions, in this case educational strategies or methods, that systematically vary from one another. They then test the comparative effectiveness of the different methods or approaches. If such approaches could be joined with a broad set of information to be provided clients, specifically education concerning treatment, disease development, and education with respect to disease burden, then we would have a much more complete understanding of the success of genetic counseling as an educational encounter and ideas on how to improve it.

How Useful Is the Information Provided in Counseling?

While clients may acquire medical and genetic information in counseling, one can ask how useful the clients find this information to be. The reason for asking this question is that the information the professional thinks the client should know may be only partly related to the decision- or problem-solving calculus the client employs. If this is the case, the professional may be able to provide information that is more useful. If there is a poor fit, however, between the professional knowledge provided and the knowledge clients use to make decisions, then the efficacy of counseling is compromised.

There has been comparatively limited research on how useful clients find genetic counseling information to be. Most information on this issue has been acquired indirectly in examining client decision-making processes subsequent to counseling. A few studies have examined the questions clients bring to counseling and the extent to which these questions are answered,[8] as well as difficulties clients experience in making decisions after counseling.[9]

In reviewing relevant studies, several impressions emerge. The first is that in some studies a sizable number of clients are uncertain about what to do reproductively after counseling; sometimes almost as many are uncertain after as before counseling. The sources of this uncertainty are not clear. These studies suggest that the information provided in counseling apparently is sometimes incomplete in terms of its utility to clients for making reproductive decisions. In several studies of the client decision-making process after counseling, some specific recurring problems or difficulties have emerged, including translating a recurrence risk statistic into a personal risk assessment, lack of availability or sources for information and support on how to make a decision, as well as conflict with relatives over an acceptable decision. Of course, while genetic counseling clients may be looking for certainty regarding their reproduction, medical genetic science and technology cannot always provide such certainty. Accordingly, while one might expect there to be some normal level of difficulty in using genetic counseling information, research could suggest ways in which existing medical and genetic information could be made more useful.

Given the paucity of studies on this issue, it seems useful to consider two types of research not present in the genetic counseling literature. First, research is needed that identifies the problems and difficulties clients routinely experience in making decisions after counseling, as well as the success-ful strategies and tactics they employ in overcoming these difficulties. To accomplish this, one might identify groups of individuals or couples who experienced few difficulties, or who experienced some, but yet were able to overcome them. By studying such individuals, an appreciation of the prob-lems and difficulties associated with using genetic counseling information, as well as their possible solutions, could be developed. Based on such informa-tion, one could then develop alternative genetic counseling approaches or strategies. Using intervention model research, these different approaches could be systematically assessed regarding their effectiveness.

If such research were undertaken and proved successful, it could go a long way toward making the word "counseling" as important as the word "genetic" in the activity called "genetic counseling."[10] While this could improve the effectiveness of genetic counseling, it could also have significant manpower training implications, as changes may be required in the training of genetic counselors as well as the amount of time allotted for genetic coun-seling activities. Nevertheless, findings from such research could also expand the current model of genetic counseling from a largely professionally defined agenda model to a professional- and client-defined agenda, hopefully increas-ing the utility of medical and genetic knowledge for client decision making subsequent to counseling.

Client Psychological Responses to Counseling

There have been comparatively few studies on the psychological responses of clients to counseling and screening, and this makes it difficult to draw any firm conclusions. Several studies of various types of screening programs have included an assortment of psychological measures.[11] To the degree that one can generalize from these studies, the impression one obtains is that people who are identified as carriers and are counseled do not appear to suffer either significant short- or long-term psychological harm or disturbance. It is not possible at this time to summarize conclusions about the psychological impact of counseling in nonscreening situations, because it has not been ade-quately researched.

Studies of the psychological impact of counseling are potentially useful in understanding why clients respond to counseling as they do, and in addition, such studies could provide useful information for improving the process of counseling. Such studies are difficult, of course, to conduct. They require the use of carefully selected and validated psychological measures and their repeated application to a panel of clients, often couples, over time. Of particular interest in such studies would be an assessment of possible "iatrogenic" consequences of counseling. More specifically, if clients experience unusual or abnormal psychological consequences, such as depression, anxiety, or self-stigmatization, then counseling could be altered or adjusted to reduce such consequences. A better understanding of client psychological responses to counseling would be important in providing the empirical base for developing the "counseling" aspect of genetic counseling. In the absence of such studies, discussion of the psychological consequences of counseling remains largely speculative, and genetic counseling remains more an educational than a counseling activity.

Reproductive Behavior of Clients

In contrast to studies on client psychological responses to counseling, there is a substantial body of research on client reproductive intentions and behavior after counseling. In large part these studies are observational in design, describing client intentions and behavior after counseling. Rarely has a study been conducted with a comparison group. Thus it is possible to describe change and stability in reproductive intentions and behavior from before to after counseling, but difficult to attribute these intentions and behavior specifically to genetic counseling.

A number of recent reviews have summarized the extensive literature on the reproductive behavior of genetic counseling clients.[12] In large part the research on client reproductive behavior after counseling suggests three major conclusions: (1) there is little change in reproductive plans and behavior from before to after counseling for many clients; (2) a major predictor of client reproductive behavior after counseling are their reproductive intentions prior to counseling, sometimes interpreted as the clients' desire for a child; and (3) both the clients' interpretation of their risk (but not necessarily their actual numeric risk) and their perception of how burdensome a child with a genetic disease would be, financially and on the family, are significant predictors of client reproductive intentions and behavior after counseling. Efforts have been made to develop statistical predictive models of client reproductive behavior, based on such factors.[13] Some such models do have significantly high predictive value. They do not, however, describe the process by which clients arrive at such decisions, information that would be useful for better understanding how to improve the counseling process.

Since the mid-1970s the prescriptive literature on genetic counseling has focused on and emphasized counseling as a largely educational activity, enabling clients to make more informed reproductive decisions. Curiously, during this same period the majority of studies on counseling have focused more on client reproductive planning and behavior than on education. Studies focusing on the processes by which clients arrive at their decisions, and not just the decision itself, could serve to better inform the counseling process.

Impact of Counseling and Screening on Couples and Families

There has been considerable discussion in the professional genetic counseling literature about the potential negative impact on couples of learning one's own or one's spouse's carrier status, or of being identified at risk to have a child with a genetic disorder. However, there has been almost no research on indicators of family disruption subsequent to genetic counseling and screening. A few early studies on genetic counseling did examine divorce rates among couples several years after counseling. Although there was some indication of elevated rates, the studies did not permit the drawing of strong conclusions. The Stamatoyannopoulous study in Greece did suggest that knowledge of carrier status, as provided to the families in that culture, was quite stigmatizing, enough to keep families from sharing the information with other families so as not to affect the mate selection process.[14] Exactly how knowledge of carrier status affected established family units is not clear from this study, nor, for that matter, from any other study to date.

Research on the impact of counseling and screening on families, if undertaken, should be designed to tap not just the most severe indicator of disruption, divorce, but should also look at other indicators, such as separation, desertion/abandonment, spouse abuse, and levels of marital satisfaction/dissatisfaction. Such studies, of course, are difficult to execute and are expensive. They require longitudinal follow-up of couples, possibly over a period of several years, extensive measurement "intrusions" into the lives of couples, a procedure that could itself affect the dependent variables being studied, and, of course, comparison groups. Moreover, any such study would have to take into account not only normally prevailing rates of divorce, separation, and marital satisfaction/dissatisfaction, but they would have to allow for elevated rates in families that experience reproductive problems in general, and in particular elevated rates in families that have a child with a birth defect or genetic disorder.

If family disruption problems are related to couples acquiring genetic information, this would be important information. The direct impact on the genetic counseling or screening session might be limited. Couples, however, could be made aware of the availability of sources of professional, as well as lay support groups, to assist them with such problems. Studies of genetic counseling to date suggest that referrals to such services are relatively rare and that genetic services, particularly those in tertiary-care settings, tend not to be "networked" strongly into either professional, lay, or self-help support systems for such problems.[15]

Summary

While much has been learned over the past twenty years about genetic counseling, screening, and client responses to such services, much remains to be learned. The Human Genome Project will both provide new opportunities for and add to the complexity of conducting research on these topics. It is likely that with expanded understanding of the role of genes in more diseases and disorders, the availability of and demand for both genetic screening and counseling will increase. It is also likely that screening and counseling will be provided in more settings, and not just medical settings, but nonmedical settings such as worksites and possibly schools as well. Given such developments, we could expand significantly the list of questions that Hecht and Holmes raised and also elaborate on the "consumer's view" issues studied by Leonard and colleagues.

The five questions identified above could be viewed as providing one priority listing of questions about genetic counseling and screening, which, if answered more fully than at present, could significantly improve clinical practice as well as public policy on applied human genetics. The decision to allocate a percent of the total dollars designated for the Human Genome Project to studying, among other things, questions of the kind listed above hopefully will mean that answers to such questions will continue to be forthcoming.

NOTES

1. F. Hecht and L. Holmes, "What We Don't Know about Genetic Counseling (letter)" *New England Journal of Medicine* 287(9):464–5, 1972.
2. C. G. Leonard, Chase, and B. Childs, "Genetic Counseling: A Consumer's View," *New England Journal of Medicine,* 287(9):433–9, 1972.
3. S. Kessler, "Psychological Aspects of Genetic Counseling," *American Journal of Medical Genetics* 34:340–53, 1989; and P. Frets and M. Niermeijer, "Reproductive Planning after Genetic Counselling: A Perspective from the Last Decade," *Clinical Genetics* 38:295–306, 1990.
4. B. Wilford and N. Fost, "The Cystic Fibrosis Gene: Medical and Social Implications for Heterozygote Selection," *Journal of the American Medical Association* 263(20):2777–83, 1990.

5. C. Epstein, et. al., "Genetic Counseling," *American Journal of Human Genetics* 27:240–3, 1975.
6. Kessler, 1989; and Frets and Niermeijer, 1990.
7. J. Sorenson, J. Swazey, and N. Scotch, *Reproductive Pasts: Reproductive Futures,* Birth Defects Original Articles Series, Vol. XVII, No. 4, 1981, Alan R. Liss, New York, 194 pp.
8. Frets and Niermeijer, 1990.
9. Ibid.
10. Sorenson, et al., 1981, pp. 131–44.
11. T. Holtzman, *Proceed with Caution,* The Johns Hopkins University Press, Baltimore, MD, 1989, 303 pp.
12. Kessler, 1989; and Frets and Niermeijer, 1990.
13. P. Frets, et. al., "Model Identifying the Reproductive Decision after Genetic Counseling," *American Journal of Medical Genetics,* 35:503–9, 1990.
14. G. Stamatoyannopoulos, "Problems of Screening and Counseling in the Hemoglobinopathies," in *Birth Defects, Proceeding of the Fourth International Conference,* Vienna, 1973, edited by A. G. Motulsky and F. Ebling, Exerpta Medica, Amsterdam, pp. 268–76.
15. Sorenson et al., 1981.

Legal and
Ethical Frontiers

13

The Potential Impact
of the Human Genome Project
on Procreative Liberty

John A. Robertson

Procreative liberty—the freedom to make decisions to beget, bear, and rear offspring—is a deeply held value and a central component of most theories of personal autonomy. In discussing the legal status of procreative liberty, however, the legal right to procreate and the legal right to avoid procreation must be distinguished.[1]

While some restrictions based on age and marital status, and to a much lesser degree mental competence, exist, it is fair to say that married couples have a fundamental right to reproduce by coital means. That is, the state would have to demonstrate a compelling need and the absence of less restrictive alternatives to limit the number, the timing, or the fact of coital reproduction.[2] It would seem to follow that married couples have a right to employ a range of noncoital means to overcome infertility, though some persons disagree as to whether this right includes the right to make and enforce surrogate-mother contracts.

It is also well-established that all persons—married or single—have a right to avoid reproduction. Whether or not they have a right to engage in sexual relations (fornication and adultery laws have not been declared unconstitutional by the Supreme Court), the Supreme Court has upheld the right of access to contraceptives and to terminate pregnancies up until viability.[3] The right in question is a right against government interference with contraception and abortion and thus does not include the right to have government fund abortions or fund family-planning programs that include abortion counselling.[4] Even if *Roe* v. *Wade* is reversed, and the right to terminate pregnancy loses its constitutional status, many states will probably continue to recognize a large degree of freedom over decisions to avoid procreation.

The rapid increase in knowledge of human genetics, which the Human Genome Project will accelerate, raises many important questions about the scope and meaning of procreative liberty. These questions arise because genetic information has the potential to increase procreative freedom by producing knowledge that expands reproductive options. At the same time, such expanded options will increase pressure for reproductive responsibility and may threaten reproductive freedom by requiring persons to acquire and make use of genetic knowledge in their reproductive decisions.

Given the strong presumption against government mandating or preventing persons from reproducing, the most salient questions concerning the impact of the Human Genome Project on procreative liberty concern whether individuals will have access to and be free to use the genetic information that they find material to their own reproductive plans.[5] The question of access concerns whether physicians must apprise patients of genetic tests or otherwise make such tests available. The question of use concerns whether the state may bar individuals from using genetic information in preconceptual, preimplantation, or prenatal screening, and then acting on the results. After these issues are discussed, this chapter will briefly address possible roles for government in relation to procreative choices based on genetic information.

In assessing these questions, several types of genetic information must be kept in mind. One type concerns information of carrier or at-risk status for serious genetic disease, for example, Tay-Sachs disease and sickle cell anemia, about which there is little disagreement regarding severity or undesirability. A second type concerns information about diseases that have variable expressivity, from very severe to relatively mild, such as cystic fibrosis. A third type of information concerns genes that may predispose toward but do not ineluctably produce serious disease, such as genes that increase the risk that one will develop heart disease, cancer, depression, or diabetes. Finally, information may develop concerning the genetic basis for desirable traits.

Information about Carrier Status

Any reasonable concept of procreative freedom should include the right of persons to know whether they and their mates are at a higher than ordinary risk for genetic disease, so they may decide whether to reproduce with that partner or remain childless, adopt, use donor gametes, or conceive and screen the pregnancy for the at-risk condition.

While carrier detection and prenatal screening programs arouse ambivalence, and in some cases opposition, few persons would limit the right of an individual or couple to know their carrier status in deciding on reproduction. Indeed, a state ban on obtaining such information would no doubt infringe a couple's right not to reproduce—a right now firmly established in Supreme Court precedents on abortion and contraception.[6] If a couple is free to not reproduce, they should also have a derivative right against states preventing access to information material to that decision. Thus the state would have limited power to prevent couples from learning their carrier status.

Currently, the main policy issues in carrier screening do not involve governmental efforts to prevent people from learning their carrier status or to require mandatory carrier screening programs. Rather, the issues concern the obligation of private-sector actors to inform persons of the availability of carrier tests of uncertain reliability and scope. That is, are physicians under a legal obligation to tell persons of tests so that they may learn their carrier status and take action on the basis of it, even if the test cannot give couples 100 percent certainty that they are not carriers of the trait in question?

The legal issues here are traditional issues of informed consent and malpractice in the special context of two problems: (1) the inability of existing tests to definitively inform persons of their status, and (2) the legality of testing for conditions or predispositions that only some persons might want to know.

The first problem is illustrated by carrier screening for cystic fibrosis (CF). A carrier test is available for mutations causing 75 percent of cases of CF, which can screen out half the couples at risk. Must physicians routinely inform persons of childbearing age of this test? Will they be held liable if they fail to do so and the couple has an affected child, which they would have not conceived or would have aborted if they had been informed of such a test?

Given that a National Institutes of Health (NIH) Workshop and other authoritative bodies have recommended against routine carrier screening for CF at the present time, the case against routine testing may seem clear. Yet legal concepts of informed consent, which view the question from the perspective of the patient who did want the information, may lead to a different result. There is always the chance that courts will independently assess the risks and benefits of testing and conclude that the benefits are sufficient to require testing, even if well-designed programs have not yet been developed.[7] Such a result is most likely in a jurisdiction that has adopted a reasonable person standard for informed consent, which focuses on the relevance of information to the patient's decision, rather than on physician practice or the physician's assessment of the need for the information.[8] If they were informed of the availability of the test, some couples might be very willing to have their carrier status for CF tested, even though only 80 percent of the leading CF mutations can now be identified.

The result of this situation is that CF carrier screening is likely to be routinely offered to many patients before the infrastructure for effective screening recommended by experts is in place. Some patients will benefit from the information, while others will experience the errors and confusion about which the experts warned. In light of this situation, the best public policy would be to proceed as quickly as possible to create the educational and counseling infrastructure essential for private-sector-induced population screening for CF. Those efforts cannot await the development of the more accurate tests that the NIH Workshop viewed as a necessary precondition for population screening, because some routine carrier screening will be introduced by physicians who are keen on avoiding legal liability. Ironically, the vagaries of tort liability may lead to a more rapid introduction of safe and effective testing procedures than if the uncertainties and distortions of tort incentives did not exist.

The problems with making population screening for CF generally available are likely to be repeated as other carrier tests emerge. Once a carrier test is available, many persons will find it in their interest to have the test, even if all the conditions for safe and effective population screening have not been established. The product of many individual decisions by patients and physicians may thus lead to the widespread use of a test that could not be justified independently as a matter of social policy. Given this reality, public policy makers should assume that carrier screening for CF and other diseases will routinely occur as tests develop and should concentrate on designing the education and counseling infrastructure necessary to make carrier screening safe and effective for all individuals concerned.

The second question—whether physicians should inform of carrier screening for genes that predispose but do not necessarily produce serious disease—will also depend on how important such information is to the reproductive decisions of couples. It would appear less important than in the case of CF, but as genetic information develops, interest in knowing one's carrier status for a wide range of predisposing genes might also develop.

Prenatal Screening for Genetic Disease

Carrier testing or other indications might lead couples to seek prenatal testing to determine whether their fetus is affected, in order that they might then end the pregnancy. The Human Genome Project will increase the conditions for which prenatal diagnosis might be sought. At the same time, prenatal screening techniques are being pushed back earlier and earlier in the pregnancy and may eventually be possible on the basis of a test of maternal serum early in the pregnancy. This will make prenatal screening easier and lead to pressures to make it more routine.

Two general types of legal, ethical, and social problems will arise in the intersection of genetics and prenatal screening. One is a private-sector issue of whether physicians are obligated to inform persons about prenatal screening tests, so that patients might make a decision about whether they wish to proceed with a pregnancy. Informed consent and malpractice considerations will come into play here, as well as the limits of the physician's role when he or she disagrees or has moral objections to a choice made by the patient.

A second general issue concerns whether society may set limits on a couple's access to prenatal information to prevent choices being made on the basis of genetic information that society deems unacceptable. The latter question is likely to be more sharply debated as prenatal screening for genes that predispose to certain conditions is sought, or as abortion laws become more restrictive.

Limits on Abortion for Genetic Indications

While most people accept prenatal diagnosis and abortion for serious genetic disease, there is much wider opposition to prenatal diagnosis and termination of pregnancy for what are perceived as less serious or trivial genetic defects. Abortion for sex selection is usually frowned upon, and many physicians have qualms about aborting a fetus with phenylketonuria (PKU) or even CF, because treatment or wide variability in those diseases exists. The reluctance to screen and abort will increase as genes that predispose for cancer, heart disease, alcoholism, depression, diabetes, and other conditions are identified. This reluctance will exist even if screening can be done on maternal serum and abortion occurs early in the first trimester.

As long as the premises of *Roe* v. *Wade* remain intact, however, a woman would have the legal right to terminate a pregnancy on the basis of such genetic information if she so desires. If a woman might abort because the pregnancy is unwanted, an inquiry into her reasons for finding the pregnancy unwanted would be barred. She would have the legal right to abort to avoid having offspring with genes that predispose to disease.

Given this legal status for abortion, legislation could not limit abortion or prenatal testing for indications that some would find trivial or unimportant, such as a genetic predisposition to heart disease or alcoholism. As long as *Roe* v. *Wade*'s recognition of a woman's right to terminate pregnancy remains intact, the right to abort for any reason seems to be constitutionally protected. *Roe* makes the pregnant woman the final arbiter of whether she will undergo the burdens of pregnancy, with her reasons or motives for that judgment not subject to legal challenge. If a woman can end the pregnancy of a healthy or normal fetus because the pregnancy is unwanted, it would follow that she could end the pregnancy because the fetus had a genetic characteristic that made continuation of the pregnancy unwelcome. She is the final judge of whether a particular medical or social condition, whether CF, PKU, or even gender, makes continuation of the pregnancy unacceptable to her.[9]

Similarly, physicians could have an obligation to inform patients of the availability of such prenatal tests, even if they are opposed to using them. In that case they would have a right not to give the test or do the abortion, but they cannot deprive the patient of information that would lead the patient to seek those services elsewhere, unless they are functioning in a federally funded program that has made refusal to discuss abortion options a condition of participation.[10]

Ethically, one may question the ethics of aborting a pregnancy because of a less serious or variably expressed genetic disease or, for that matter, aborting a healthy fetus because the pregnancy is unwanted. The ethical debate will depend on one's evaluation of the moral status of the fetus at the particular stages of development at which abortion occurs, the importance of a woman's choice in deciding whether to continue gestation, and one's perceptions of genetic disease.[11] But if abortion is accepted for unwanted pregnancy, there is no easily discernible ground for distinguishing among the reasons for abortion under *Roe*. Of course, physicians remain free to refuse to perform abortion for reasons that they find unacceptable and, with appropriate notice, many even withhold the results of prenatal diagnostic tests to prevent abortion from occurring.

Prenatal Diagnosis and Abortion
if Roe v. Wade Is Reversed

If changes in the composition of the Supreme Court lead to the reversal of *Roe* v. *Wade,* the status of abortion after prenatal screening will depend on the specifics of legislation in particular states. While some states will not change the existing array of rights, some states will pass restrictive abortion laws. However, many restrictive states are likely to allow abortion for threats to health, rape, or incest or in cases of "fetal abnormality" or "severe fetal defect."

The question posed under these statutes is what counts as a "fetal abnormality" or "severe fetal defect"? Unless specific conditions or criteria are listed, such laws could be attacked as unconstitutionally vague. If they are to be made more specific, what conditions should be included? In this situation, abortion to avoid offspring with genes that predispose to heart disease, cancer, and so forth is not likely to qualify. A harder question is whether conditions such as PKU or CF would qualify, since they may be treated or have widely varied expression. Precise answers to this question will have to await future developments. At that time, however, direct grappling with what is "normal," "abnormal," and a "defect" serious enough to justify abortion will have to occur. Indeed, this need illustrates well how genetic information will force us to confront the meaning of "normal."

Even if abortion for many genetic conditions were banned, it would not follow that prenatal diagnostic tests for those conditions could also be banned. A state's ban on prenatal testing to prevent persons from going to other states to obtain abortions would be subject to constitutional attack as an interference with the right of interstate travel.[12] Also, knowing the genetic status of the fetus before birth would be helpful in managing the pregnancy and birth process. Thus a ban on all prenatal testing would be vulnerable to constitutional attack on several grounds, even if abortion itself were prohibited.

Preimplantation Genetic Diagnosis and Screening of Embryos

Preimplantation screening of embryos is still highly experimental, with work in humans only beginning. As a technical matter, such screening requires the ability to isolate embryos, to sample blastomeres, and then to assay their genetic structure.[13] Although isolation of embryos is feasible by in vitro fertilization (IVF), much more work remains to be done on sampling blastomeres and assaying cells. Polymerase chain reaction assaying of blastomeres to determine their sex has been reported in mice and humans.[14] Rapid progress in this area and application to humans can be expected soon.

Postconception, preimplantation screening of early embryos would occur as follows. Extracorporeal embryos will be produced by either IVF or uterine lavage. Fertilized eggs that have cleaved and appear likely candidates for placement in the uterus will have a cell or blastomere removed, which will be subjected to genetic analysis. With rapid techniques, such as polymerase chain reaction amplification of DNA, a quick genetic assessment can occur. Alternatively, the embryo could be frozen while the genetic diagnosis takes place. If the genetic tests are negative, the embryo, either fresh or after thawing, can be placed in the uterus in the hope that pregnancy will occur. If the tests are positive, the embryo need not be transferred, but can be discarded or allowed to deteriorate. The advantage of preimplantation testing is that it avoids the need for screening and abortion after pregnancy has occurred. In addition, it opens the door eventually to gene therapy on the embryo.

Even when perfected, however, preimplantation genetic screening of embryos, owing to the difficulty of isolating embryos and inducing pregnancy thereafter, is likely to appeal to only a minority of women. Women otherwise undergoing IVF to treat infertility might choose to have their embryos screened, if the screening does not reduce the chances of achieving pregnancy coitally. For example, women who are against abortion generally but at risk for CF might find preimplantation screening more acceptable.[15] Or women who have had the unfortunate experience of aborting a fetus after amniocentesis might choose this method. But IVF is sufficiently arduous and expensive that even couples who have a one in four chance of offspring with a serious genetic disease will not always seek preimplantation screening.

Assuming that blastomere biopsy can occur without hurting the embryos or reducing the pregnancy rate, strong ethical objection to preimplantation screening may be expected. The objections arise from differing views about embryo status,[16] from concerns that preimplantation genetic screening will open the door to germline genetic therapy,[17] and from more general concerns about manipulating nature.[18] Some persons also object to the deliberate production of embryos for research that will be necessary to perfect the techniques in question.[19]

Despite these concerns, selection of preembryos on genetic grounds appears to be ethically acceptable.[20] The goal of avoiding the birth of offspring with severe genetic handicaps is part of procreative liberty and parental discretion. Pursuing this goal prior to implantation enables at-risk couples to avoid the birth of a handicapped child without having to undergo prenatal diagnosis and abortion.

The means employed to achieve this goal—preimplantation genetic diagnosis and nontransfer of embryos—is also ethically acceptable. Because embryos are so rudimentary in development, they are not generally viewed as having interests or rights. Thus they have no right to be placed in a uterus and may be discarded if they carry the gene for serious disease. Indeed, it is preferable to discard at the embryonic stage rather than abort fetuses that are more fully developed.

Finally, the fear of slippery slopes is not sufficient to render unethical a practice that is otherwise ethically acceptable. Deselection on grounds of serious genetic disease does make possible deselection on less serious grounds. It also opens the door to germline gene therapy and is no doubt a necessary step if positive genetic engineering of other offspring traits is to occur. Yet there is no showing that any of these possibilities is so clearly unacceptable or so likely to occur, that the potential for such outcomes justifies denying at-risk couples the benefits that genetic selection of embryos makes possible.

Other ethical issues with preimplantation genetic diagnosis are issues that arise with any new medical technology. They concern the ethics of offering a procedure whose safety and efficacy has not been clearly established. The danger is that patients will be led into thinking that the procedure offers a realistic possibility of real net benefit, when it is still highly experimental. Close attention to the ethics of human experimentation and informed consent are necessary to avoid these problems.

The Role of Government

The discussion above has focused on the right of couples to have or use genetic information in making reproductive decisions *free* of government interference. Thus a major role of government is a negative or noninterfering role, with the corresponding duty to refrain from barring persons from obtaining or using genetic information in reproduction.

Consideration should also be given to facilitative policies to make sure that persons of childbearing age are aware of genetic options. Greater support for education and genetic counseling would be appropriate in this regard.[21] Also appropriate might be requirements that physicians inform couples and pregnant women of screening options, as some states have done. Such a policy would respect procreative freedom by assuring that people have access to the information they need to make reproductive choices.

A governmental role that went beyond regulating or facilitating the provision of genetic information and mandated that persons learn their carrier and fetal status and then act on it to prevent having offspring with "unacceptable" genetic characteristics cannot be justified. Some persons, for example, might argue that all persons applying for marriage or drivers' licenses be screened for heterozygote genetic characteristics, so that they may make marriage and reproductive decisions with this information in mind. Indeed, some persons might even ban marriage between two carriers or require that they not be free to reproduce unless they screen embryos and fetuses for unacceptable genetic traits. Finally, some persons might also favor prenatal screening of all pregnancies, with abortion required or encouraged if the results are positive.[22]

Merely stating such proposals for a more intrusive governmental role makes the case against them apparent. Coercive governmental intrusion in private, reproductive matters is a frightening infringement of personal, marital, and reproductive freedom. It recalls the eugenic policies of the Nazis and the American sterilization laws of the 1920s that were so widely abused. The notion of governmental intervention to certify that persons may reproduce or that a woman could continue a pregnancy to term is simply intolerable to our scheme of basic freedom as we now conceive it, for the most basic freedoms are involved in the reproductive decisions to which genetic information would be relevant.[23]

Conclusion

This brief survey of reproductive uses of genetic knowledge at the carrier and prenatal stages shows some of the moral and legal issues that are likely to arise as new genetic information flows out of the Human Genome Project. In the final analysis, the new genetics cannot escape our strong normative commitment to procreative liberty. Because genetic knowledge is relevant to the reproductive choices of many persons, it should be available to the widest extent possible. By the same token, it should also remain a matter of personal choice. Individuals, not government, should decide how to use genetic knowledge in reproduction.

NOTES

1. J. A. Robertson, "Procreative Liberty and the Control of Conception, Pregnancy and Childbirth," *Virginia Law Review* 69:405, 401–10, 1983.
2. J. A. Robertson, "Embryos, Families and Procreative Liberty: The Legal Structure of the New Reproduction," *Southern California Law Review* 59:939, 954–64, 1986a.
3. Robertson, 1983, p. 405.
4. See *Maher* v. *Roe,* 432 U.S. 464 (1977); *Harris* v. *McRae,* 448 U.S. 297 (1980); *Rust* v. *Sullivan,* 111 S. Ct. (1990).
5. J. A. Robertson, "Procreative Liberty and Human Genetics," *Emory Law Journal* 39:697, 1990.
6. *Roe* v. *Wade,* 410 U.S. 113 (1973); *Griswold* v. *Connecticut,* 381 U.S. 479 (1965).
7. See, for example, *Helling* v. *Carey,* 83 Wash. 2d 514, 519 P.2d 981 (1974) (practice of ophthalmologists does not prevent court from independently assessing the need for a diagnostic test).
8. See, for example, *Canterbury* v. *Spence,* 464 F. 2d 772 (D.C. Cir.), cert. denied, 409 U.S. 1064 (1972) (patient must be informed of information that would be material to a person in his situation).
9. In fact, abortions for purposes of gender determination appear to be very rare. A post-*Webster* restrictive abortion law enacted in 1989 in Pennsylvania contained a provision prohibiting abortion for sex selection. 1990 Pa. Legis. Serv., Act 1989-

64, S.B. No. 369, § 3204(c) (Purdon). The American Civil Liberties Union suit challenging that law did not attack the sex selection provision because such abortions are too rare to be of concern.

10. See *Rust* v. *Sullivan,* 111 S. Ct. 1759 (1990). Private programs may also set limits on the services or the counseling they provide, if state law permits them to do so. For example, state law could permit a private clinic or program to provide no information that could lead to abortion, including information about prenatal genetic tests. Of course, patients should be informed in advance of such limitations.

11. Obviously different persons will weigh these considerations differently, depending on their values, life experiences, and many other factors.

12. See *Shapiro* v. *Thompson,* 394 U.S. 618 (1969).

13. J. L. Simpson, S. A. Carson, J. E. Buster, and S. Elias, "Future Horizons in Prenatal Genetic Diagnosis: Preimplantation Diagnosis and Non-invasive Screening 6," in: *Human Prenatal Diagnosis,* 2nd ed., edited by K. Filkins and J. F. Russo, Marcel Dekker, New York, 1990, pp. 547–7.

14. A. H. Handyside, R. J. Penketh, R.M.L. Winston, J. K. Pattinson, et al., "Biopsy of Human Preimplantation Embryos and Sexing by DNA Amplification," *Lancet,* 1:347, 1989. See also "Scientists Identify Sex of 3-Day-Old Embryos," *New York Times,* April 19, 1990, p. A19, col. 1.

15. This assumes that some women who would object to abortion of a fetus with CF would not have the same qualms about discarding a preimplantation embryo with CF, because of the great difference in their biological development.

16. See, for example, Ethics Committee of the American Fertility Society, "Ethical Considerations of the New Reproductive Technologies," *Fertility and Sterility* 46(Suppl. 1):29S, 1986.

17. L. Walters, "The Ethics of Human Gene Therapy," *Nature* 320:225, 227, 1986.

18. See, for example, Robertson, 1986a; A. Huxley, *Brave New World,* Harper & Roe, New York 1932, pp. 3–5, 10–4.

19. To perfect preimplantation diagnosis of embryos will necessitate a great deal of embryo research, itself a very controversial issue. For a survey of concerns that arise with embryo research, see Robertson, "Embryo Research," *University of Western Ontario Law Review* 24:15, 1986b.

20. J. A. Robertson, "Ethical and Legal Issues in Preimplantation Genetic Screening," *Fertility and Sterility* 57:1, 1992.

21. These are the measures that Wilfond and Fost recommend as necessary to implement mass carrier screening for CF. See Wilfond and Fost, "The Cystic Fibrosis Gene: Medical and Social Implications for Heterozygote Dectection," *Journal of the American Medical Association* 263:2777, 1990.

22. No one, to my knowledge, is recommending any such policy. But there are periodic proposals from state legislators to sterilize women on welfare after they have a certain number of children. Others have also recommended that women with AIDS be urged to have abortion.

23. However, in a very extreme case of pressing need, either severe under- or overpopulation, perhaps some limitation on procreative liberty could be justified. See, for example, "Comment, Legal Analysis and Population Control: The Problem of Coercion," *Harvard Law Review* 84:1856, 1971.

14

Patent Rights in the Human Genome Project

Rebecca S. Eisenberg

The various research efforts that comprise the Human Genome Project will inevitably both draw on and yield a multitude of patentable inventions. The broad subject matter of the patent laws potentially reaches every phase of the Genome Project, from the discovery of new research technologies, such as techniques and equipment for DNA sequencing, through the ultimate development of new products, such as screening tests for genetically transmitted diseases. Even bits and pieces of the human genome itself may be, and sometimes have been, patented.[1] Nor does the fact that the public is paying for the Genome Project through federal funding mean that the public may freely enjoy the fruits of that research. Quite the contrary, existing law not only permits, but affirmatively encourages, patenting and private commercial exploitation of inventions made in the course of the Genome Project.

Nonetheless, the prospect of private ownership of knowledge emanating from the Genome Project has provoked controversy. The National Research Council Committee on Mapping and Sequencing the Human Genome concluded in its 1988 study that "human genome sequences should be a public trust" not subject to the intellectual property laws,[2] while the Office of Technology Assessment's 1988 report on the Genome Project suggested that federal agencies and Congress should instead promote early filing of patent applications followed by prompt release of data.[3] By early 1992, the controversy had focused on the filing by the National Institutes of Health (NIH) of patent applications on some 2,375 partial cDNA sequences identified in its laboratories, along with the as-yet unidentified full genes and gene products to which they correspond.

Given the magnitude of resources invested in the Human Genome Project in both the public and private sectors and the tremendous potential benefits to be reaped from this research, the role of the patent laws in this area deserves careful thought. Ideally, patent law should promote the progress of research and product development and the dissemination of research results. This chapter clarifies some of the implications of patent law for the Genome Project, with a view to identifying problems that might interfere with the smooth operation of the patent system in this context.

Summary of Existing Law and Policy

A U.S. patent confers the exclusive right to make, use, or sell the patented invention in the United States for seventeen years from the date the patent issues.[4] During this term, the patent holder has the right to prevent anyone from using the invention–even an innocent infringer who develops the same invention independently,[5] and even a subsequent inventor who improves on the basic invention to such a degree that the improvement itself earns a patent.[6] In exchange for these broad exclusive rights, the inventor must disclose the invention to the public in terms that enable others who are "skilled in the art" to make and use it.[7] So long as the disclosure requirement is satisfied, it is not necessary for the patent applicant to have actually made a tangible embodiment of the invention in the laboratory. Judicial decisions characterize the disclosure as the "quid pro quo" of the patent monopoly.[8] In order to obtain a patent, the applicant must first contribute "a measure of worthwhile knowledge to the public storehouse."[9] A patent consists of a written description of the invention and of how to make and use it, often accompanied by figures or drawings, and one or more "claims" specifying exactly what it is that others may not make, use, or sell.[10]

It is a fundamental axiom of patent law that one may patent only that which is new. Section 101 of the Patent Act defines as patentable subject matter "any new and useful process, machine, manufacture, or composition of matter, or any new and useful improvement thereof."[11] The Supreme Court has construed this language broadly to include "anything under the sun that is made by man," including genetically engineered living organisms.[12] Although products of nature may not be patented as such, patents have been issued on such products in human-altered form. For example, bacteria or chemicals that are newly isolated and purified may be patented in an isolated and purified state if they exist in nature only in an impure state.[13] Consistent with these principles, patents have been issued on proteins and DNA sequences from the human genome that have been purified and isolated through human intervention.[14] The requirement that a patentable invention be "useful" excludes from protection certain scientific discoveries that, although interesting as a subject of further research, cannot yet be used for any practical human purpose.[15] Some critics of the NIH patent applications on partial cDNA sequences have argued that these inventions fail to satisfy this utility requirement, but few patents are rejected on this basis. Patents have been issued for DNA sequences that code for useful proteins[16] or that serve as probes to detect genetic susceptibility to disease.[17]

In addition to being new and useful, an invention must satisfy the further statutory requirement of nonobviousness[18] to be patentable. One may not obtain a patent by disclosing an "invention" that is already available to the public, whether because it was previously known or because it is readily discoverable through obvious advances over the prior art. As the prior art in a field grows, the nonobviousness requirement makes more and more subsequent discoveries unpatentable.

For the past decade federal patent policy has actively encouraged the patenting and commercialization of inventions made in the course of federally sponsored research.[19] Legislation enacted in 1980 requires small businesses and nonprofit organizations, including universities,[20] to report promptly to the funding agency any potentially patentable inventions made in the course of sponsored research.[21] The statute permits recipient institutions to retain title to their inventions if they agree to file patent applications in a timely manner and to ensure that the inventions are utilized.[22] Patent holders may either exploit their patents themselves or license someone else to exploit them, but if they do neither, the government reserves "march-in" rights to grant licenses in order to ensure practical application of the inventions.[23]

The underlying premise of this statutory scheme is that the public benefits more from inventions made in the course of federally sponsored research when those inventions are patented and exploited commercially in the private sector.[24] Inventions left in the public domain are presumed to languish in government and university archives rather than to be freely and widely exploited. Note that this vision of the role of patents is different from the time-honored justification for the patent system as a means of promoting investment in research to make new inventions. The traditional conception of the role of patents concedes that patents actually restrict dissemination of the inventions they cover, but nonetheless justifies the creation of patent monopolies as a means of inducing firms to undertake the necessary research to make such inventions in the first place. In the context of federally sponsored inventions, the government is not offering patent rights as an *ex ante* incentive to stimulate future inventions, but rather is insisting on patent protection *ex post* for inventions that have already been made. The justification is that existing patent rights facilitate the commercial dissemination of products embodying the results of prior research, not that the prospect of future patent rights will stimulate future research. While this philosophy may seem counterintuitive to scientists whose norms enjoin them to make their research results freely available to the public,[25] for the time being it seems to be firmly entrenched in federal law and policy.

Impact on the Genome Project

What is the likely impact on the Genome Project of a policy that encourages patenting of federally funded inventions? Will such a policy ultimately promote or impede full utilization of knowledge gained through this research? A thorough response to this question would require a broad reexamination of the effectiveness of the patent system as a whole. Rather than taking on such a daunting task in these brief remarks, I will instead assume, along with federal policy makers, that in many contexts the patent system is an effective means of promoting widespread dissemination of new discoveries through commercial channels. The question I will pose is whether there are particular reasons to question its effectiveness in this specific context. I shall first analyze the likely impact of patent incentives on the conduct of the research itself and then consider the effectiveness of patent law in promoting full commercial utilization of knowledge generated in the Genome Project.

Impact on the Conduct of Research

One reason for concern that a broad federal policy in favor of patenting research results might be unsuitable for the Genome Project is that the Project is a long-term, ongoing research project in which continued progress will depend on prompt, unfettered access to prior discoveries. To the extent that patent law creates barriers to such access, it may impede progress.

Perhaps the biggest danger that patenting presents to progress in the Genome Project is that researchers seeking to preserve patent rights will defer publication of their findings and thereby retard the dissemination of new knowledge. While the patent system ultimately compels disclosure of whatever information is necessary in order to make and use a patented invention,[26] disclosure through the U.S. patent system does not occur until a patent issues,[27] which is likely to be years after a discovery is made. Academic researchers who want to earn recognition for their discoveries in the scientific community may wish to publish their inventions in journals at an earlier date, but early disclosure is risky for those who plan to seek patent protection. A scientist who discloses an invention in a publication before applying for a patent forfeits patent protection immediately in most of the world, although U.S. patent law allows filing of an application up to one year thereafter.[28]

The solution to this dilemma might seem to be for scientists to file patent applications promptly and then publish their research results,[29] but there are problems with this strategy. For one thing, research may yield publishable results before it yields a patentable invention.[30] In this situation publication of early results could prevent patenting of later-developed inventions emanating from the same research if the publication makes the subsequent inventions "obvious."[31] Another risk is that publication of a narrow form of an invention may preclude patenting of a more broadly claimed invention at a later date.[32] Thus a fully informed scientist who wants both patent protection and scientific recognition may publish her discoveries later than she would if she had no interest in patent protection. Delays in publication of research results pertaining to patented inventions could be minimized by accelerating the progress of patent applications through the Patent and Trademark Office, but this problem is hardly unique to the Genome Project.

One might question how many scientists participating in the Genome Project will actually delay publishing their work to any significant degree because of patent considerations. For commercial scientists, the possibility of obtaining patent rights might make publication more likely than it would be if they had to rely on secrecy to preserve intellectual property rights in their discoveries.[33] As for scientists in universitites, a recent empirical study found that academic scientists receiving funds from biotechnology companies both patented and published their research results at a higher rate than their colleagues who did not receive such funding,[34] suggesting that patent considerations need not interfere unduly with publication. But it stands to reason that, faced with the risk of losing patent rights through premature publication, some scientists will publish research results later than they would if they had no concern about patents.

A potentially more harmful consequence of patenting discoveries made in the course of the Genome Project is that patent holders could restrict access to these discoveries in ways that impede subsequent research. A patent holder normally has complete discretion whether to exploit the invention as a monopolist, to license it on an exclusive or nonexclusive basis, or to suppress the invention entirely. Such broad exclusive rights could potentially retard scientific progress, particularly in the case of patents on research technologies developed early on for use in later stages of the Genome Project.

But present law does provide exceptions to the general rule of private control over access to publicly funded inventions. These exceptions might be invoked to retain some measure of public control over the results of the Genome Project. Specifically, the statute allows the funding agency to exercise "march-in" rights to grant or compel the granting of licenses to federally funded inventions if it determines that such action is necessary "to achieve practical application of the subject invention."[35] So far, the meaning of this language has not been tested in litigation.

At the very least, this provision should permit mandatory licensing of inventions that would otherwise go unused. It is less clear whether it permits such licensing in the case of inventions that are being used exclusively by the patent holder. For example, a grant recipient that discovered and patented a group of partial cDNA sequences might want to keep the sequences to itself in order to give its own researchers an advantage over competitors in finding (and possibly patenting) the corresponding full genes and gene products. In such a situation, the patent holder might try to avoid the exercise of march-in rights by arguing that it has achieved "practical application" of the inventions in its own laboratories, and thereby satisfied the statute, without having to make the inventions available to others. On the other hand, if the practical effect of such exclusivity is to restrict utilization of the invention to a significant degree, a compelling case could be made for the exercise of march-in rights. As articulated in the patent statute, the purpose of allowing grant recipients to retain patent rights is "to promote the utilization of [federally funded] inventions," to be sure that such inventions "are used in a manner to promote free competition and enterprise," "to promote the commercialization and public availability of [such] inventions," and to "protect the public against nonuse or unreasonable use of inventions."[36] Hoarding of patented research intermediaries for the exclusive use of patent holders, to the detriment of research in other laboratories, flies in the face of these policies.

Two additional exceptions in the statute appear potentially applicable in this context, although so far funding agencies have construed both of them narrowly. One provides that a funding agency retains a paid-up, nonexclusive license "to practice or have practiced for or on behalf of the United States [the federally funded invention] throughout the world."[37] On its face, this language arguably permits agencies to authorize subsequent grant recipients to use the inventions of prior grant recipients without payment of royalties on the theory that they are practicing the inventions "on behalf of the United States" when they use them in federally sponsored research. So far, however, agencies have viewed this provision as limited to situations where the government wants to obtain a patented product for strictly governmental purposes from someone other than the patent holder.[38]

The other statutory exception allows an agency to prevent recipients of research funds from retaining title to inventions "in exceptional circumstances when it is determined by the agency that restriction or elimination of the right to retain title to any subject invention will better promote the policy and objectives of this chapter."[39] The language of the statute and regulations promulgated thereunder suggest that this exception should be used sparingly.[40] The statute presupposes that its objectives are ordinarily best achieved by allowing fund recipients to retain patent rights. In order to show "exceptional circumstances," it would seem necessary at the very least to distinguish the research results at issue from other inventions made in the course of federally funded research, and to explain why the patent system is less likely to promote utilization and dissemination of inventions in this context than in other contexts.

Finally, apart from these statutory provisions, the courts have recognized a limited, nonstatutory exception to a patent holder's exclusive rights for the use of a patented invention in pure, noncommercial research.[41] It is difficult to determine the scope of this exception with clarity because, although many courts have acknowledged its existence in principle, almost none have actually applied it to the facts of the cases before them to excuse otherwise infringing activity. A recent judicial decision characterized the experimental use defense as "truly narrow,"[42] and it seems unlikely to provide significant protection for researchers until its terms are clarified through further caselaw or legislation.

A statutory "experimental use" defense to permit the use of patented inventions in the Genome Project would not be without precedent[43] and has much to recommend it as a matter of policy.[44] One difficulty with a royalty-free experimental use exemption for subsequent researchers is that in the case of a patent on research technology, researchers are the only consumers of the invention. A rule that exempts these consumers from infringement liability entirely would eliminate the patent holder's profits and effectively eviscerate the incentive for commercialization offered by the patent. Nor would such an exemption ordinarily be necessary in order to make patented research technology available to researchers: holders of patents on such technology will generally view researchers as potential customers and will want to extend licenses to them in order to collect royalties. But in some situations, particularly when the subsequent researcher is doing further research in competition with the patent holder, the patent holder might have more to lose by allowing the use to proceed than it stands to gain by collecting royalties from the researcher. In this context an experimental use defense may be necessary for the research to proceed, although it might be appropriate to compel the subsequent to user to pay royalties to the patent holder.[45] In effect, this would amount to a compulsory license in favor of subsequent researchers rather than an exemption from infringement liability.

If the only purpose of an experimental use exemption is to compel the holders of patents on federally funded inventions to extend licenses to subsequent researchers, this objective might possibly be achieved without any new legislation through the exercise of march-in rights by funding agencies. A statutory experimental use exemption might nonetheless be useful in that it could clarify ambiguity as to the scope of the exemption under existing case law, obviate the need for funding agencies to go through cumbersome procedures to exercise march-in rights,[46] and ensure researchers access to inventions whether or not they were made with federal funds.

The problem of patent seekers deferring disclosure of new findings may be more intractable than the problem of patent holders restricting access to their inventions.

Impact on Practical Utilization of New Knowledge

The argument for patent protection is less threatening to the research community when the invention at issue is a medical product or process for use by the public rather than research technology for use in the laboratory.[47] For example, mapping and sequencing a gene of interest may lead to the development of a screening test for a genetic disease or to the production of a useful protein through recombinant DNA technology. Patent protection for such inventions is less troubling and more compelling than it is for inventions that are primarily of interest to researchers. Making a new invention available to a market of ordinary consumers is a function traditionally performed by commercial firms in our economy, even when there is a strong public interest in widespread availability of the invention at low cost, as in the case of a new drug. It stands to reason that commercial firms will be more willing to perform this function if their profits are protected by patent rights than if they can recover only a competitive rate of return. Moreover, if an invention has a significant market outside the research community, the patent owner might focus its enforcement of the patent on the more lucrative nonresearch market and leave the occasional laboratory user alone. Indeed, the patent owner may welcome the use of its invention by researchers in the hope that they will develop new uses for the patented technology and thereby open up new markets for the patent owner to exploit.

Assuming that patent protection is necessary to encourage firms to make these discoveries available commercially, one might ask whether current law provides adequate protection. It is helpful to distinguish different types of inventions emanating from the Genome Project and to consider the availability and consequences of patent protection for different categories of invention separately.

So far, new biotechnology products have typically emerged from the following sequence of discoveries. First, a naturally occurring protein is identified as having medical significance, perhaps because its absence causes disease symptoms in individuals who fail to produce adequate quantities of it in their own systems. The protein is then isolated and purified from natural sources, allowing it to be characterized and perhaps making it available in small quantities for therapeutic purposes. Next, the gene for the protein is cloned, disclosing the DNA sequence encoding the protein and allowing its production in larger quantities through recombinant DNA technology. The recombinant protein is then clinically tested and eventually displaces the natural protein in sales for therapeutic purposes.

As the Genome Project proceeds, the direction of discovery is reversed in some cases as researchers look for genetic causes of diseases whose biochemical mechanisms are not yet understood. In this "reverse genetics,"[48] the first step is to locate the gene on a specific region of a particular chromosome through genetic linkage studies using DNA markers. At this point it may be possible to use the markers to develop a genetic screening test to identify carriers of the disease gene. Depending on the distance between the closest known markers on either side of the gene, it may be necessary to undertake biochemical studies of differences in gene product in the tissue of affected and unaffected individuals in order to find the actual gene of interest. Eventually, the disease gene itself is found, cloned, and sequenced, and the protein associated with the disease is identified.

The NIH patent applications on partial cDNA sequences reveal yet another discovery path, in which the first step is to sequence portions of randomly selected clones from a cDNA library representing the genes expressed in a given tissue sample. These partial sequences, called expressed sequence tags or ESTs, may then be used to isolate the full gene, the function of which is subsequently determined.

The foregoing trails of discovery yield a series of candidates for patent protection: purified, naturally occurring proteins; protein purification processes; DNA sequences coding for proteins of interest; DNA sequence markers for use in genetic screening tests; DNA sequences whose function has not yet been determined; recombinant vectors and host cells; cloning processes; recombinant proteins; processes for obtaining recombinant proteins from host cells; and processes for administering natural or recombinant proteins in the treatment of disease. Some of these discoveries are end products for sale to consumers, and some are processes for using these end products. Others may be thought of as manufacturing processes yielding end products or starting materials for use in manufacturing processes, or research intermediaries for use in developing products and processes of more immediate commercial interest.

From the patent holder's standpoint, generally the most effective commercial protection is offered by a patent on the end product that is sold to consumers. Such a patent is superior to a patent on a process of using the end product in that it is not limited to any particular use of the product, and it is superior to a patent on the manufacturing process in that it is not limited to any particular method of production. Somewhat less effective is a patent on starting materials for use in making the end product. Such a patent offers no protection against use of the patented starting materials abroad and then importing the unpatented end product into the United States for sales in competition with the patent holder.[49] Weaker still is a patent on products or processes used only during product development, since an injunction against infringement of such a patent would not serve to keep a competitor off the market so long as the use is not continuing.

Problems with Process Claims. Before the Supreme Court upheld the patentability of living organisms in *Diamond v. Chakrabarty,* process patents were the mainstay of patent protection for biotechnology inventions. Patents were issued and their validity upheld on claims to processes using bacteria to treat sewage,[50] or to produce chemicals[51] or drugs,[52] although the bacteria used in these processes were generally assumed to be unpatentable. Today it is clear that recombinant microorganisms and host cells may be patented in their own right, assuming they are new, useful, and nonobvious. But there are still advantages to obtaining patent rights in processes using these biological materials to make end products, particularly if the end products themselves are not patentable.

The most significant advantage is that a process patent offers superior protection against competing imports of the end product. The difference is due to peculiarities in the definition of infringement arising from recent amendments to the patent statute to fortify process patent protection. Prior to 1988, the commercial effectiveness of process patents was largely limited to use of the claimed process in the United States. A competitor could use the patented process abroad and import the unpatented products of the process into the United States for sale without infringement liability, although in some cases the patent holder had a limited remedy against importation of the products under the Tariff Act.[53] This gap in process patent protection was largely remedied with passage of the Omnibus Trade and Competitiveness Act of 1988,[54] which included a provision expanding the definition of patent infringement to include importing into the United States or selling or using within the United States a product made by a process patented in the United States.[55]

But the statutory change failed to provide comparable protection against imports to holders of product patents on essential starting materials such as recombinant organisms. As a result, the holder of a patent on a recombinant organism useful to make a valuable protein has no remedy against a competitor who uses the organism abroad and imports the end product into the United States.[56] In order to obtain such a remedy, it is necessary to obtain a patent on the process of using the recombinant organism to produce the protein.

In recent years, however, the Patent and Trademark Office has taken a restrictive view of the patentability of such biotechnology processes. Relying on a 1985 decision of the Court of Appeals for the Federal Circuit,[57] the Patent and Trademark Office has rejected as obvious patent claims to conventional processes using novel and nonobvious starting materials to produce novel and nonobvious end products.[58] Legislation is currently pending in Congress that, if passed, would provide that process patent protection shall not be denied on grounds of obviousness for a process of making or using a patentable product.[59] In the meantime, firms that use standard processes to express recombinant proteins in novel recombinant organisms may find that they are unable to patent the processes and must instead try to patent the recombinant starting materials or protein end products.

Even apart from questions of patentability, there are problems with detection and proof of infringement of process patents. Unless the patented process is the only means of making an unpatented product, it may be difficult to tell whether a competitor used the patented method or some other, noninfringing method. Moreover, even if the patented process is initially the only means of making the product, a process patent offers no protection against competitors who develop noninfringing means of making the product in the future. Patents on processes for using, as opposed to making, unpatented products suffer from a similar problem. A patent on a process of using an unpatented product for a certain purpose, such as to treat a particular disease, is difficult to enforce if the product has other, noninfringing, uses because it may be hard to monitor just what purchasers are doing with the product. Thus even where process patent protection is available, the commercial monopoly it offers may be narrow, difficult to enforce, and short-lived.

Problems with Protein Claims. As a general rule, a patent on an end product provides more powerful protection than a patent on either starting materials or processes for making or using the end product. In the biotechnology industry the end product is often a protein. Claims to the protein product itself would avoid the difficulties cited above for monitoring and enforcing process patent claims and would be less vulnerable to changes in technology that allow manufacture of the product through other means.

On the other hand, the validity of patent claims to proteins may be more vulnerable to challenge than that of process claims. One ground for challenge is that proteins are unpatentable products of nature. The proteins that are likely targets for commercial production through recombinant DNA technology for the most part exist and perform important functions in nature. So far patent holders have avoided this difficulty by limiting their claims so as not to cover the naturally occurring product. For example, the claims may specify a degree of purity for the protein that is not duplicated in nature without human intervention. Such limitations on the scope of the claim, while possibly avoiding a "product of nature" rejection, may also narrow the scope of the patent to a degree that allows significant competition beyond the reach of the patent monopoly. For example, a competitor might avoid infringement by selling a product that is either more or less pure than that claimed in the patent.

A competitor might also be able to avoid the scope of the patent monopoly by varying the chemical structure of the protein if it is possible to do so without destroying its function. Sometimes minor changes in amino acid sequence do not interfere with the biological activity of a protein. Some patent applicants have attempted through careful claim drafting to include minor variations in amino acid sequence within the scope of the patent, but a recent decision of the Court of Appeals for the Federal Circuit calls into question the effectiveness of this practice.[60]

Another limitation to patents on proteins is suggested in a recent Federal Circuit opinion in a lawsuit brought by the holder of a patent on purified Factor VIII:C against a firm that was making recombinant Factor VIII:C.[61] The court noted that although the patent claims on purified Factor VIII:C obtained from blood plasma literally covered the recombinant product, the recombinant manufacturer might nonetheless be able to avoid infringement liability under the rarely invoked "reverse doctrine of equivalents"[62] if it could show sufficient differences in specific activities and purity between the recombinant product and the plasma-derived product.[63]

A further problem with protein claims is that the protein may have been known and characterized in the literature before the gene encoding it was cloned. In this situation the protein may have become unpatentable on grounds of prior knowledge, use, or publication,[64] or it may even be claimed in a prior patent[65] that presents an obstacle to subsequent innovators who wish to produce the protein through recombinant DNA technology.

Even previously unknown proteins may become unpatentable on grounds of obviousness as the Genome Project progresses. Indeed, once the complete DNA sequence of the human genome is known, it might be argued that all proteins encoded in that known sequence have become obvious.

Problems with DNA Sequence Claims. The foregoing discussion reveals a number of circumstances in which a patent applicant's best hope may be a product claim on a DNA sequence or recombinants incorporating that sequence, although such a claim offers less effective commercial protection than either a product claim on the protein end product or a process claim on the method of producing the protein. The protein itself may be unpatentable because it was previously known or even patented, and the process may be unpatentable because it is obvious. A DNA sequence may also be the most likely target for a patent claim where the invention consists of identifying a gene that is associated with an inherited disease or a DNA marker in close proximity to that gene rather than cloning a gene for the purpose of expressing a protein. In this situation the DNA sequence itself may be thought of as an end product, useful in a genetic screening test, rather than as a starting material in making a protein.

Like proteins, DNA sequences are arguably unpatentable products of nature. So far the Patent and Trademark Office and the courts have allowed patent applicants and patent holders to get past this hurdle by limiting their claims to an "isolated and purified" DNA sequence,[66] or to a human-engineered recombinant vector or host cell incorporating the sequence.[67]

Claims to chromosomal DNA markers not coding for any identified gene product might face a further hurdle in the requirement that a patented invention be useful. The fact that markers are useful to scientists seeking to map and sequence the human genome may not be sufficient evidence of utility to make them patentable.[68] On the other hand, a marker that is sufficiently close to a disease gene to be useful in screening for carriers should certainly satisfy the utility test. In its patent applications on partial cDNAs, for example, NIH asserts that the sequences are useful for forensic identification and tissue typing.

As cloning procedures become more routine, it is likely that DNA sequences coding for known proteins will be deemed obvious, especially when the amino acid sequence for the target protein is known. It is also possible that knowledge of partial cDNA sequences will make obvious the corresponding full genes. So far, in assessing the obviousness of patent claims to DNA sequences the courts have focused on the obviousness of the process of cloning and identifying the gene of interest, upholding claims to sequences that were difficult to isolate.[69] As the Genome Project progresses, greater experience will inevitably make more DNA sequences obvious under this approach.

Even when they are patentable, claims to DNA sequences may have limited commercial value. Where DNA sequences are, in effect, starting materials for making desired proteins, patents on the DNA sequences offer less effective commercial protection, particularly against foreign competition, than either patents on the proteins themselves or process patents on the method of making the proteins.[70] Patent claims to partial gene sequences used only as research intermediaries to find the full genes offer even less effective protection.

Another potential problem with claims to DNA sequences as starting materials has to do with the scope of the claims in light of the degeneracy of the genetic code. Numerous substitutions may be made in a nucleotide sequence without changing the amino acid sequence of the gene product. Do these substitutions remove the altered sequence from the scope of the patent claim? The answer certainly should be no, but the matter is not entirely free from doubt.

The Court of Appeals for the Federal Circuit touched on this issue in its recent decision in the case of *Amgen Inc. v. Chugai Pharmaceutical Co.*[71] One of the patents at issue in that case included a broadly worded claim to any purified and isolated DNA sequence "encoding a polypeptide having an amino acid sequence sufficiently duplicative of that of erythropoietin to allow possession of [biological properties of erythropoietin]."[72] The court held this claim invalid for lack of an adequate disclosure of how to make more than a few of the many "analog genes" that would fall within the scope of the claim.[73] The court was careful to note, however, that it did not "intend to imply that generic claims to genetic sequences cannot be valid where they are of a scope appropriate to the invention disclosed by an applicant."[74]

Although the opinion is not free from ambiguity on this point, a reasonable reading would allow patent holders to claim DNA sequences in sufficiently broad terms to include, at the very least, analogous sequences that do not alter the gene product. It is well understood among geneticists that the genetic code is redundant in that certain amino acids may be encoded by any of a number of "codons" of three nucleotides. These interchangeable codons have long been identified, and the substitution of one for another in a DNA sequence would be obvious to anyone skilled in the field. To permit competitors to avoid the patent through such inconsequential substitutions would be manifestly unfair and contrary to established principles of patent law.[75] By contrast, changes in DNA sequence that alter the gene product in ways that do not change its biological activity are far more difficult to predict. Thus a sensible reading of the *Amgen* decision would prevent generic claiming of DNA sequences when the variations claimed yield different gene products, but not when the variations in the sequence amount to substitutions of recognized equivalents that yield the same gene product.

Conclusion

The enormous potential commercial implications of knowledge gained through the Genome Project make it inevitable that some of that knowledge will be patented. Federal law encourages this result by promoting the patenting of inventions made in the course of federally sponsored research. This federal policy rests on the assumption that patenting helps to ensure widespread availability of the results of sponsored research through commercial dissemination. The wisdom of this policy rests in large part on the empirical validity of this assumption. If the assumption is valid, then there may be cause for concern that patent law provides inadequate protection for some biotechnology inventions at this time. Of particular concern are limitations on the patentability of biotechnology processes and possible limits on the scope of protection available for proteins and DNA sequences. On the other hand, the current federal policy may cause scientists to retard dissemination of new findings within the scientific community and to restrict access to new inventions among scientists, all to the detriment of progress in the Genome Project. If it turns out that patent protection is not necessary in order to ensure widespread public dissemination of these inventions, this may be a big price to pay for little or nothing in the way of social benefits.

Given the importance of funding from private industry to biomedical research today, one should not assume that patent protection for the results of the Genome Project is unnecessary. The small amounts of federal funding projected for the Project cannot be expected to displace private funding in this area. Moreover, even if public funding were sufficient to cover mapping and sequencing efforts, lack of patent rights might undermine incentives for the private sector to support subsequent research to put this information to practical use. In the meantime, obstacles to progress in the Genome Project as a result of patent rights might be addressed through more limited reforms without calling into question the wisdom of promoting patent protection. Problems of interim secrecy could be ameliorated by decreasing delays in the prosecution of biotechnology patent applications in the Patent and Trademark Office. Concerns about restrictions on the use of patented inventions in subsequent research might be met through the exercise of march-in rights by funding agencies or through enactment of an experimental use exemption to infringement liability in this context. These sorts of reforms might help minimize friction between the patent system and traditional scientific norms without jeopardizing incentives for private support of research in this area.

ACKNOWLEDGMENTS

I am grateful to Roberta Morris for helpful comments on these remarks in draft form, and to John Ogilvie for able research assistance.

NOTES

1. See Eisenberg, "Patenting the Human Genome," *Emory Law Journal* 39:721, 1990.
2. Committee on Mapping and Sequencing the Human Genome of the National Research Council, *Mapping and Sequencing the Human Genome* 99–100, National Academy Press, Washington, D.C., 1988 [hereinafter National Research Council Committee Report]; see also, V. McKusick, "Mapping and Sequencing the Human Genome, *New England Journal of Medicine* 320:910, 912, 1989.
3. U.S. Congress, Office of Technology Assessment, *Mapping Our Genes—Genome Projects: How Big, How Fast?* 166–69, U.S. Government Printing Office, Washington, D.C., 1988 [hereinafter OTA Report].
4. 35 U.S.C.A. § 154 (West 1984).
5. 35 U.S.C.A. § 271 (West 1984 and Supp. 1990).
6. In this situation neither patent holder would be able to use the improvement without first obtaining a license from the other.
7. 35 U.S.C.A. § 112 (West 1984). Section 112 also requires that the inventor disclose the "best mode" for making and using the invention known to the inventor at the time the application is filed.
8. See *United States* v. *Dubilier Condenser Corp.,* 289 U.S. 178, 186–87 (1933); *Grant* v. *Raymond,* 31 U.S. (6 Pet:) 218, 247 (1832).
9. *Application of Argoudelis,* 434 F.2d 1390, 1394 (C.C.P.A. 1970) (Baldwin, J., concurring).
10. 35 U.S.C.A. § 112 (West 1984).
11. 35 U.S.C.A. § 101 (West 1984). The requirement that an invention be "new" is elaborated upon in 35 U.S.C.A. § 102 (West 1984).
12. *Diamond* v. *Chakrabarty,* 447 U.S. 303, 309 (1980) (quoting S. Rep. No. 1979, 82d Cong., 2d Sess. 5 (1952) and H.R. Rep. No. 1923, 82d Cong., 2d Sess. 6 (1952)).
13. See, for example, *In re Bergy,* 563 F.2d 1031 (C.C.P.A. 1977), *vacated,* 438 U.S. 902 (1978), *on remand,* 596 F.2d 952 (C.C.P.A.), *cert. granted sub nom. Parker* v. *Bergy,* 444 U.S. 924 (1979), *vacated and remanded with instructions to dismiss as moot sub nom. Diamond* v. *Chakrabarty,* 444 U.S. 1028 (1980) (biologically pure culture of microorganism); *In re* Bergstrom, 427 F.2d 1394 (C.C.P.A. 1970) (pure prostaglandins); *Parke-Davis & Co.* v. *H.K. Mulford & Co.,* 189 F. 95, 103

(S.D.N.Y. 1911), *rev'd in part on other grounds,* 196 F. 496 (2d Cir. 1912) (purified adrenalin composition); *Kuehmsted v. Farbenfabriken,* 179 F. 701 (7th Cir. 1910), *cert. denied,* 220 U.S. 622 (1911) (acetyl salicylic acid)

14. See Eisenberg, 1990, p. 721, note 4, and sources cited therein.
15. *Brenner v. Manson,* 383 U.S. 519, 535 (1966).
16. See, for example, U.S. Patent 4,994,371 (Feb. 19, 1991) (isolated DNA sequence encoding human factor IX); U.S. Patent 4,970,161 (Nov. 13, 1990) (recombinant plasmids and animal cells incorporating DNA sequence encoding human inter-feron-gamma).
17. See, for example, U.S. Patent 4,965,189 (Oct. 23, 1990) (oligonucleotide probes for use in assaying for type I diabetes mellitus through hybridization with human DNA).
18. 35 U.S.C.A. § 103 (West 1984); see also *Graham v. John Deere Co.,* 383 U.S. 1, 13–19 (1966) (interpreting § 103 as a codification of case law requiring an inde-pendent determination of nonobvious as a prerequisite to patentability).
19. 35 U.S.C.A. § § 200 et seq. (West 1984 and Supp. 1990). Section 200 provides: "It is the policy and objective of the Congress to use the patent system to promote the utilization of inventions arising from federally supported research or develop-ment; . . . to promote the commercialization and public availability of inventions made in the United States by United States industry and labor. . . ." See generally Rosenfeld, "Sharing of Research Results in a Federally Sponsored Gene Mapping Project," *Rutgers Computer and Technical Law Journal* 14:311, 1988.
20. 35 U.S.C.A. § 201(i) (West 1984). The same policy was extended to other govern-ment contractors—that is, large businesses—in a 1983 Presidential Memoran-dum. "President's Memorandum to the Heads of Executive Departments and Agencies on Government Patent Policy," *Weekly Comp. Presidential Documents* 19:252, Feb. 18, 1983.
21. 35 U.S.C.A. § 202(c)(1) (West Supp. 1990). Failure to report an invention within a reasonable time results in forfeiture of title to the invention to the government. Ibid.
22. 35 U.S.C.A. §§ 202(a), (c) (West Supp. 1990). If the recipient does not elect to retain title, the agency may step in and pursue patent rights itself. 35 U.S.C.A. § 202(c) (2) (West Supp. 1990). Alternatively, the individual inventor may retain rights to the invention. 35 U.S.C.A. § 202(d) (West 1984).
23. 35 U.S.C.A. § 203 (West 1984 and Supp. 1990)
24. 35 U.S.C.A. § 201 (West 1984); Rosenfeld, 1988, p. 316; OTA Report, 1988, p. 167.
25. See R. Eisenberg, "Proprietary Rights and the Norms of Science in Biotechnology Research," *Yale Law Journal* 97:177, 1987.
26. 35 U.S.C.A. § 112 (West 1984).
27. 35 U.S.C.A. § 122, 11 (West 1984). If the inventor also seeks foreign patent pro-tection, the contents of the patent application will be disclosed eighteen months after it is filed.
28. 35 U.S.C.A.§ 102(b) (West 1984).
29. This is the recommended approach in the OTA Report, 1988.
30. See Eisenberg, 1987, p. 216.
31. 35 U.S.C.A. § 103 (West 1984). See, for example, In re O'Farrell, 853 F.2d 894 (Fed. Cir. 1988).

32. See *In re Ruscetta,* 255 F.2d 687 (C.C.P.A. 1958).
33. See Eisenberg, 1987, pp. 214–16.
34. Blumenthal et al., "University-Industry Research Relationships in Biotechnology: Implications for the University" *Science* 232:1361, 1986.
35. 35 U.S.C.A. § 203(a) (West Supp. 1990). The government may also exercise march-in rights if it determines that "health or safety needs . . . are not reasonably satisfied" by the patent holder, or that "such . . . action is necessary to meet requirements for public use specified by Federal regulations and such requirements are not reasonably satisfied" by the patent holder. 35 U.S.C.A. § 203(b),(c) (West 1984). See also 35 U.S.C.A. § 203(d), which permits the exercise of march-in rights in order to ensure that preference is given to U.S. industry in exploiting the invention.
36. 35 U.S.C.A. § 200 (West 1984).
37. 35 U.S.C.A. § 202(c)(4) (West Supp. 1990).
38. See Rosenfeld, 1988, pp. 333–7.
39. 35 U.S.C.A. § 202(a)(ii) (West Supp. 1990).
40. In order to invoke this exception, the agency must comply with procedural requirements set forth in the statute and in Department of Commerce regulations. 35 U.S.C.A. § 202(b) (West Supp. 1990); 37 C.F.R. § 401.3 (1990). The agency must file with the Secretary of Commerce a written determination that exceptional circumstances exist including an analysis justifying the determination.
41. This exception is analyzed at length in R. Eisenberg, "Patents and the Progress of Science: Exclusive Rights and Experimental Use," *University of Chicago Law Review* 56:1017, 1989.
42. *Roche Prods., Inc.* v. *Bolar Pharmaceutical Co.,* 733 F.2d 858, 863 (Fed. Cir. 1984), *cert. denied sub nom. Bolar Pharmaceutical Co.* v. *Roche Prods., Inc.,* 469 U.S. 856 (1984).
43. Congress has enacted experimental use exemptions in the Plant Variety Protection Act of 1970, 7 U.S.C.A. § 2544 and the Semiconductor Chip Protection Act of 1984, 17 U.S.C.A. § 906(a).
44. See generally Eisenberg, 1989.
45. Ibid., pp. 1075–8.
46. See 37 C.F.R. § 401.6 (1990).
47. Of course, there is no bright line between discoveries that are useful in medical applications and discoveries that are useful for scientists working in the laboratory. An example of a patented technology that finds markets among both medical consumers and researchers is polymerase chain reaction (PCR). PCR, which is the subject of several patents owned by Cetus Corporation, for example, U.S. Patent No. 4,965,188 (October 1990), U.S. Patent No. 4,683,195 (July 1987), U.S. Patent No. 4,683,202 (July 1987), permits selective amplification of extremely small amounts of DNA by as much as a billionfold. "Cetus Corp. Obtains Fifth Patent on Its PCR Gene Amplification Technology," *Business Wire, Inc.,* October 24, 1990 (Lexis wire service).
48. OTA Report 1988, pp. 61–2.
49. See notes 53–6 and accompanying text.
50. *Cameron Septic Tank Co.* v. *Village of Saratoga Springs,* 159 F. 453, 462 (2d Cir. 1908), *cert. denied,* 209 U.S. 548 (1908); *City of Milwaukee* v. *Activated Sludge, Inc.* 69 F.2d 577, 582–83 (7th Cir. 1934), *cert. denied,* 293 U.S. 576 (1934).

51. *Guaranty Trust Co.* v. *Union Solvents Corp.,* 54 F.2d 400, 403, 410 (D. Del. 1931), *aff'd* 61 F.2d 1041 (3d Cir. 1932), *cert. denied,* 288 U.S. 614 (1933).

52. *In re Mancy,* 499 F.2d 1289 (C.C.P.A. 1974). See Wegner, "Patent Protection for Novel Microorganisms Useful for the Preparation of Known Products," *International Review of Indus. Prop. and Copyright Law* 5:285 (1974); Note, "Microorganisms and the Patent Office: To Deposit or Not to Deposit, That Is the Question," *Fordham Law Review* 52:592, 595–6, 1984.

53. § 337(a) of the Tariff Act of 1930, ch. 497, 46 Stat. 590, codified at 19 U.S.C.A. § 1337(a) (West 1980 and Supp. 1991).

54. Pub. L. No. 100–418, 102 Stat. 1107, codified in pertinent part at 35 U.S.C.A. § 271(g) (West Supp. 1990).

55. 35 U.S.C.A. § 271(g) (West Supp. 1990).

56. See *Amgen, Inc.* v. *U.S. Int'l Trade Comm'n,* 902 F.2d 1532 (Fed. Cir. 1990); *Amgen, Inc.* v. *Chugai Pharmaceutical Co.,* 9 U.S.P.Q.2d (BNA) 1833, 1843–47 (D. Mass. 1989).

57. In re Durden, 763 F.2d 1406 (Fed. Cir. 1985).

58. See, For example, *Ex parte Kifer,* 5 U.S.P.Q.2d (BNA) 1904 (Bd. Pat. App. & Interf. 1987); *Ex parte Orser,* 14 U.S.P.Q.2d (BNA) 1987 (Bd. Pat. App. & Interf. 1989). A more recent opinion of the Federal Circuit arguably limits the reach of *In re Durden* in the case of processes for *using* (as opposed to *making*) patentable products, *In re Pleuddemann,* 910 F.2d 823 (Fed. Cir. 1990), although the opinion is not entirely coherent on this point and its implications for subsequent patent applications are therefore unclear.

59. Biotechnology Patent Protection Act of 1991, H.R. 1417, Cong. Rec. E946 (February 14, 1991) and S 654, Cong. Rec. S3284 (February 13, 1991), reproduced in *BNA Patent, Trademark and Copyright Journal* 41:436, March 21, 1991.

60. *Amgen, Inc.* v. *Chugai Pharmaceutical Co.,* 927 F.2d 1200, 1212–14 (Fed. Cir. 1991) (holding invalid "generic" claims to any purified and isolated DNA sequence "encoding a polypeptide having an amino acid sequence sufficiently duplicative of that of erythropoietin to allow possession of [biological properties of erythropoietin]"). See discussion in text at notes 71–5.

61. *Scripps Clinic & Research Foundation* v. *Genentech Inc.,* 927 F.2d 1565 (Fed. Cir. 1991).

62. The doctrine of equivalents holds that an "accused product" may infringe a patent even though there are differences between the accused product and the literal language of the patent claims if the accused product nonetheless "performs substantially the same function in substantially the same way to obtain the same result." *Graver Tank* v. *Linde Air Products,* 339 U.S. 605 (1950). Conversely, the reverse doctrine of equivalents holds that an accused product may avoid infringement, even though it falls within the literal language of the claim, if it is "so far changed in principle from a patented article that it performs the same or a similar function in a substantially different way." Ibid., pp. 608–9.

63. *Scripps,* 927 F.2d at 1581.

64. 35 U.S.C.A. §§ 102(a),(b) (West 1984).

65. Ibid.

66. See, For example, *Amgen, Inc.* v. *Chugai Pharmaceutical Co. Ltd.,* 927 F.2d 1200, 1206 (Fed. Cir. 1991) ("It is important to recognize that neither Fritsch nor Lin invented EPO or the EPO gene. The subject matter of claim 2 was the novel *purified and isolated* sequence which codes for EPO. . . .") (emphasis in original).

67. See, For example, U.S. Patent No. 4,970,161 (November 13, 1990), claiming plasmids and recombinant animal cells incorporating chromosomal DNA sequence coding for human interferon-gamma.

68. *Brenner* v. *Manson,* 383 U.S. 519 (1966). See text accompanying note 15.

69. See generally Eisenberg, 1990, pp. 729–36. The Federal Circuit followed this approach without analyzing its propriety in *Amgen, Inc.* v. *Chugai Pharmaceutical Co. Ltd.,* 927 F.2d 1200 (Fed. Cir. 1991).

70. See notes 53–6 and accompanying text.

71. 927 F.2d 1200 (Fed. Cir. 1991).

72. Ibid., p. 1212. See note 60.

73. Ibid., p. 1214.

74. Ibid.

75. Indeed, in the situation described in the text, the competitor would probably be liable for infringement under the doctrine of equivalents even if the patent claims did not specify that they covered such obvious substitutions. See *Graver Tank* v. *Linde Air Products,* 339 U.S. 605 (1950); see also discussion in note 62.

15

Speaking Unsmooth Things about the Human Genome Project

Thomas H. Murray

Rarely, if ever, has something gone so swiftly from prophetic warning to cliché. In just three years the claim that the Human Genome Project will raise important ethical, legal, and social issues has made that journey. The time is past to chant it like a mantra; rather, we have begun to look beneath the true but superficial cliché and to limn the specific challenges posed. We have identified a variety of issues. We also need to identify social means—formal and informal—for tracking the Genome Project, identifying new issues or significant variants of familiar ones, and assessing and choosing how we, as a political community, will deal with them.

One approach to answering this question is to inventory the various social institutions and practices by which the Genome Project can be tracked. The answer is the familiar litany of formal and informal political institutions: Congress, state legislatures, federal, state, and local courts, federal and state regulatory agencies, industrial, professional, and public-interest lobbying organizations, community, ethnic, and religious groups, and some bodies set up especially to watch the Genome Project and related phenomena such as the Recombinant DNA Advisory Committee (RAC) and the Working Group on Ethical, Legal, and Social Issues (ELSI). There are also individuals who can identify issues: scholars plying their trade, writing mostly for academic journals, and journalists, in both the print and electronic media.

Are the existing, uncoordinated set of institutions adequate to identify and deal with the issues likely to be raised by the Genome Project? Should new institutions, tailored to the need, be established? What should scholars and close observers of the project, do?

The first question, about the adequacy of existing means, is the subject of the first section of this chapter. The second section will deal with a particular new institution, the Program on Ethical, Legal, and Social Issues. Its limitations are considerable; its potential contributions, modest. Two broad issues that illustrate the kinds of challenges the Genome Project will present are the subject of the third section. Finally I will consider the roles that scholars and commentators on science and technology—that is, people like us—might play.

Is the Genome Project All That Different?

Conspiracy theorists could speculate that the Genome Project is a brilliant strategy to deflect the attention of journalists, scholars, and social critics from other issues that were *really* important, such as Middle East policy, inequities in health care, racial tensions, educational failures—choose your favorite issues. I am not prone to such theories, mostly because I doubt people are clever enough to conduct large-scale conspiracies without leaving plenty of footprints. But the speculation is useful in prompting us to ask why we are so interested at this moment in the Genome Project, and whether it is really all that important or unique.

The reason there is so intense a level of interest now is not mysterious. Bioethics has wrestled with issues in genetics from its own beginnings some thirty years ago. The questions remain intellectually challenging and humanly important. Two circumstances have given renewed impetus to that interest. First, the sheer volume of new information and new technologies promised— or threatened—by the Genome Project gives the old questions new urgency and hints that relatively novel ones will emerge. Second, the sudden availability of the largest amount of money ever to be spent on bioethics has attracted the attention of a host of scholars, educators, and others, such as television producers.

The first factor, urgent and novel questions, is a good reason for our interest. The second—the availability of money in heretofore unheard-of amounts—could be good or bad: good if it stimulates high-quality work and leads to enriched public understanding of what is at stake; bad if it is spent on mediocrity or seduces us away from equally or more significant issues for which comparable funding does not exist. Essentially the same debate goes on about science and the Genome Project, just as the wisdom of other focused scientific projects, such as the war on cancer of the 1960s and 1970s, was questioned. Probably the most useful research done under the rubric of the war on cancer was that which never lost sight of fundamental scientific questions and posed the particular research problem in the context of larger scientific issues.

More likely than not, there are other current issues that, in hindsight, will prove to be as important as or more important than the Genome Project. To accept this is simply to admit our limited ability to know what will be important in the future. It should also help put the effort to understand the ethical, legal, and social issues the Genome Project engages in the context of the full set of issues we confront.

The skeptic might ask why we need a special effort to look at such issues in the Genome Project when we have muddled along well enough with the challenges posted by other social endeavors and technological changes. We have coped with other changes in belief and technology without a comparable, organized effort. But this criticism misses the point. The question is whether and in what ways such a self-conscious program might have enabled us to handle past innovations better. The most optimistic hope is that the effort on the Genome Project will permit us to handle its many manifestations more wisely and democratically, and that such an approach may serve as a model for dealing with future scientific and technological challenges. At the very least, as long as we insist on keeping the Genome Project in context, as one of many sources of moral and social challenge, we are less likely to be blinded by dollars, or to believe that we are inventing ethical and legal questions anew.

A constructive way to begin an analysis of ethical, legal, and social issues in the Genome Project is to ask in what ways it is dissimilar to other scientific and technological movements. We are likely to find differences in degree and combination rather than differences in kind. A few such differences seem to stand out.

For one thing, contemporary genetics, of which the Genome Project is but one part, has been characterized by an extraordinarily rapid transition from basic research to practical applications. The polymerase chain reaction (PCR), for example, turned science fiction into routine practice in a few years. PCR can take DNA and use it to make virtually unlimited copies of itself, enough to enable forensic laboratories to do "DNA fingerprinting" and enough to allow embryologists to use the DNA in a single cell from a human blastocyst to determine the sex of and the presence or absence of many genetic diseases in the child that blastocyst might become. Such rapidity allows little time to ruminate over the ethical, legal, and social issues that might result.

Genetics is a science and technology of human inequality. It emphasizes the ways in which individuals and groups differ from one another. Research, for example, in the HLA antigen complex shows different frequencies among different racial and ethnic groups. This has potentially enormous social significance because certain HLA patterns are associated with humanly important outcomes such as propensities toward certain diseases. The HLA antigens are also used to match donated organs with potential recipients. African-Americans, for reasons poorly understood, are less likely to donate kidneys, more likely to need them, and less likely to get them than people of European descent. One of the reasons they are less likely to get kidneys is the use of HLA antigen matching in the allocation formula, given the relative scarcity of African-American donors. Proposals have been made to keep DNA fingerprint files separately for different groups. Such separate files might enhance the validity of DNA fingerprinting, but they might also invite discriminatory uses of the same information. These are but two of the ways genetics may divide us along lines of race and ethnicity.

In addition to emphasizing our differences, genetics emphasizes the continuity between generations, between ancestors and descendants. Parents whose children inherit genetic diseases may feel responsible for their children's affliction. Children who inherit such diseases may feel resentment toward parents. We can criticize such feelings as irrational, but our criticism ignores the importance people attach to these relationships. We need to inquire into the sources of human significance and meaning in biological inheritance.

The history of the past hundred years is replete with examples of the misuse of genetic ideas. Recent scholarship by Robert Proctor[1] and Sheila Faith Weiss,[2] for example, demonstrates the apparent ease with which genetics can become a tool in the service of social prejudices, injustice, and worse. Daniel Kevles has demonstrated the resonance eugenic ideas have found in the United States.[3] Of all the sciences, genetics is not the only one to have been misused for political purposes, but it does seem to be especially susceptible to such misuse. I suspect that a major reason for that susceptibility is the convenient and reassuring way genetics can be used to explain and justify social differences among individuals and groups. Whether this is correct is less important than recognizing the recurring phenomenon.

One more feature sets off genetics from most other scientific and technological endeavors: genetic correlations with the chracteristics of individuals have inescapable, but complex, implications for judgments about personal responsibility. We are less inclined to assign credit or blame to persons when their achievement, or illness, or misbehavior is at least in part genetically determined.

ELSI: What It Is Not and What It Is

That there is a Program on Ethical, Legal, and Social Issues in the Genome Project is a good sign, but one easily misinterpreted. First, it is important to say what such a program is not. ELSI is not a guarantee that all is well in the

Genome Project, that all moral problems will be neatly anticipated, dissected, and managed. It is ludicrous to think that a handful of scholars and clinicians who compromise the ELSI Working Group are capable of such heroic wisdom and foresight, or even to believe that the many scholars whose independent work is being funded through the ELSI program can do the same. The ELSI Working Group is not a commission or regulatory agency empowered to speak for the public or to exercise control in the public interest. Experts in medicine or ethics or law, although they may clarify issues and offer useful critiques of public policy, lack the moral and political authority to decide what ought to be done.

There are dangers here. The public, including public officials, must not be misled about what ELSI can do, lest it let down its guard. Another possibility is that by creating a core of ELSI experts, the public may feel that it lacks the knowledge or right to deal with the relevant issues. Or it may be that the experts, through friendship, familiarity, or funding, may lose their critical distance and become apologists for the scientists and enthusiasts for the technology. There is a similar phenomenon in bioethics when a nonphysician ethicist "goes native," often symbolized by donning the physician's white coat.

The ELSI component of the Genome Project originated at a congressional hearing held to discuss the Genome Project (as it was then called). I was asked to discuss the ethical issues it was likely to raise. At the end of my testimony, I urged that a portion of the energy and money appropriated for the Genome Project be devoted to study the ethical, legal, and social issues it raised. The congressmen spent much of the time with the full panel of witnesses exploring ethical issues. James Watson was also on that panel, and indicated a strong interest in ethical issues. When Watson was appointed director of NIH's genome effort, he announced his intention to devote 3 percent or more of the overall budget to ethical, legal, and social issues.

Whether the ELSI program as presently conceived will do all that its proponents hope is uncertain. But the program occupies a unique place in the history of science: it is the first major scientific initiative to include from its inception a commitment to systematically exploring the ethical, legal, and social issues it raises.

An ELSI program can make some modest contributions. It can, for example, stimulate public interest in and sophistication about such issues. It can develop materials to give the public more equal footing in the current and coming public policy debates. It can encourage attention to ethical, legal, and social issues in genetics (and other sciences and technologies) in schools and colleges. It can contribute to public education in many ways. ELSI can underwrite scholarship that identifies emerging issues, provides thoughtful analyses, and explores the advantages and disadvantages of possible policy options. ELSI, through both NIH and DOE, offers competitive grant programs for conferences and research projects.

ELSI has done at least two other things of note. The ELSI Working Group explored the issues raised by the availability of a test for carriers of cystic fibrosis (CF) a development that occurred during its brief lifetime. The Working Group was struck by the opportunity for a careful study of the evolution of CF carrier testing both for its own sake and as a precedent for other new genetic tests. It was also concerned about the possibility that the introduction of CF carrier testing would be done hastily and cause a great deal of harm. ELSI members were startled to learn that no branch of NIH was willing, at that time, to sponsor a careful study of such screening. The Working Group urged that the ELSI program and other branches of NIH collaborate to solicit and fund such pilot studies. This was done. The other potential contribution ELSI is attempting to make is to focus attention on specific issues via the creation of task forces, such as its Task Force on Genetic Information and Insurance.

For the scholars who work in or under the sponsorship of ELSI, scholarship and education are honorable and familiar endeavors. There is a third task scholars and other intellectuals can perform. It is, to use a world now out of fashion, prophecy. I do not mean crass prediction. I mean it in the sense Michael Walzer discusses in his *Interpretation and Social Criticism,* as a challenge to "the leaders, the conventions, the ritual practices of a particular society . . . in the name of values recognized and shared in that same society."[4] I will return to prophecy at the end of this chapter.

Genetic Procrusteanism and Comforting Apologies

The intense debate over intentional genetic modification of humans, such as gene therapy, seems to have distracted many people from other ways in which genetic technologies can be used to manipulate people. My worries about this are colored by my work on performance-enhancing drug use in sports with the U.S. Olympic Committee's Committee on Substance Abuse Research and Education. In that arena, where athletes are inclined to view their bodies as machines to be manipulated, where the quest for a competitive advantage is intense and often ruthless, and where the rewards of success can be astounding, technologies for self-manipulation are frequently welcomed and prized. Two biosynthetic hormones—human growth hormone and erythropoietin—have already found their way into the hands of athletes. There is no reason to believe that it will end with these two.

Some scholars take the attitude that such efforts at self-manipulation are unstoppable, unmanageable, or both.[5] Clearly, the temptations have and will come. Even before biosynthetic growth hormone was available, some parents sought it for their normal children in the hope that they would be taller and reap the benefits of this culture's favoritism toward taller adults.[6]

One of the most difficult questions we must face is whether and how to control the use of technologies that may be performance enhancing, or be desired by individuals for other reasons, but that threaten justice or other important social values. For example, when people are free to buy growth hormone for their normal children, it is likely that only the wealthy will be able to afford it. We could leave it to the market and face the prospect of adding height to the advantages that money can buy. Or we could provide access for all children. Of course, if all were a few inches taller, relative differences in height would remain, and no one would be better off except the manufacturers of growth hormone, psychotherapists (who, now that height had become a social obsession, would find a vast new clientele among the relatively shorter), and clothing fiber manufacturers, among others.

We do not know and cannot anticipate all the particular forms that the temptation to manipulate and improve ourselves and our offspring will take. But it seems clear that an argument framed primarily in terms of individual liberty will find few, and weak, reasons for restraint. We must talk about substantive conceptions of the good in particular social practices if we are to make sense of what is at stake.[7]

There are good arguments against allowing such power to manipulate humankind to fall into the hands of a few. Who has the wisdom to make such judgments? Who has the authority? Imagine this power in the hands of nativists, or the Victorians, or a committee of ethicists and lawyers. Most people can see the dangers in these scenarios. It is more difficult to show that a market-based individualism poses its own perils, yet it surely does.

In his book, *Racial Hygiene: Medicine under the Nazis,* Robert Proctor describes *apology* as one of the functions of science. Our ineluctable tendency for comforting explanations combines with the readiness of genetic accounts of individual and group differences to make a tempting brew. Genetic explanations of group differences ought to be presumptively suspect, especially if they let the powerful and comfortable explain the plight of the powerless and poor as the outcome of genetics rather than social circumstances. This is hardly a new argument, but it will have to be repeated countless times as each new candidate genetic explanation arises, and as the popular perception of the importance of genetics increases, as it almost certainly will for a time.

Conclusion

> The people say, according to Isaiah, "Speak unto us smooth things" (30:10), and that is what the professional prophets of courts and temples commonly do. It is only when these foretellings are set, as Amos first sets them, within a moral frame, when they are an occasion for indignation, when prophecies are also provocations, verbal assaults on the institutions and activities of everyday life, that they become interesting.[8]

Walzer describes the debate between the prophet Amos and his opponent Amaziah in these terms: "The claim that God is better served by scrupulous worship of himself than by just dealings with one's fellows . . . has an enduring appeal: worship is easier than justice."[9] The prophet as social critic can only succeed by invoking "the core values of his audience in a powerful and plausible way."

Good public education and good scholarly analysis on the issues raised by the Genome Project is needed. But there is also a need for prophecy in Walzer's sense of connected social criticism, criticism that is fully situated within a particular culture with a particular history, that speaks forcefully to the core values of that culture, that expresses indignation, that provokes, and that makes a people face their failure to live up to those values.

The Human Genome Project will confront our core values in many ways: our acceptance of our imperfections and mortality; our compassion; our willingness to accept responsibility for our actions; our commitment to justice. No single institution can assure that we will make the right choices when these values are at stake or even that we will recognize when they are at risk. We will need a chorus of prophets. Most will fail, either because they misunderstand our core values or cannot tell stories that persuasively connect our choices to our values. There is no reason to believe that scholars make particularly good prophets, even if the media and the public seem to think of them as such. When we see a threat to our core values, though, we must have the courage to speak unsmooth things.

NOTES

1. Robert N. Proctor, *Racial Hygiene: Medicine under the Nazis,* Harvard University Press, Cambridge, MA 1988.
2. Sheila Faith Weiss, *Race Hygiene and National Efficiency: The Eugenics of Wilhelm Schallmayer,* University of California Press, Berkeley, CA 1987.

3. Daniel J. Kevles, *In the Name of Eugenics,* Alfred A. Knopf, New York 1985.
4. Michael Walzer, *Interpretation and Social Criticism.* Harvard University Press, Cambridge, MA 1987, p. 89.
5. Norman Fost, "Banning Drugs in Sports: A Skeptical View," *Hastings Center Report* (4):5–10, 1986.
6. Thomas H. Murray, "The Growing Danger of Gene-Spliced Hormones," *Discover,* February 1987, pp. 88–92.
7. Jeffrey Stout, *Ethics after Babel: The Languages of Morals and Their Discontents,* Beacon Press, Boston, 1988.; Alasdair MacIntyre, *After Virtue,* University of Notre Dame Press, Notre Dame, IN 1981.
8. Walzer, 1987, p. 70.
9. Walzer, 1987, p. 89.

16

A National Advisory Committee on Genetic Testing and Screening

LeRoy Walters

The first part of this chapter presents the case for establishing a national advisory committee to oversee the fields of genetic testing and screening in the United States. I then discuss several possible goals for such a committee and conclude with three alternative models for the creation of a committee, two in the public sector and one in the private sector. Throughout, my indebtedness to important work accomplished by several former and existing committees will be evident. Chief among these are the President's Commission on Bioethics, the Recombinant DNA Advisory Committee (RAC) at the National Institutes of Health (NIH) and the RAC's Human Gene Therapy Subcommittee, and the Interim Licensing Authority in the United Kingdom.

A Tale of Two Technologies: Gene Therapy and Genetic Diagnosis

Two of the earliest fruits of the Human Genome Project are likely to be new tests for genetic status and gene-mediated approaches to the treatment of disease. The latter technology, often called human gene therapy, began to be discussed in the late 1960s,[1] was first seriously attempted in 1980, and was first employed in the United States in 1990. The former technology, genetic testing and screening, had its beginnings in the 1960s, long before the advent of recombinant DNA research or the initiation of the Human Genome Project.[2]

In the latter half of the 1960s and in the 1970s, the technology of genetic diagnosis diffused rapidly. The cooperative, worldwide effort to map and sequence the human genome will undoubtedly accelerate the pace at which new methods for diagnosing genetic disease, genetic predispositions, and carrier status in human beings are developed.

Although both human gene therapy and genetic testing and screening usually involve applications of recombinant DNA techniques, and although both technologies were carefully studied by the President's Commission on Bioethics,[3] the two technologies have been treated quite differently by public policy makers in the United States. By the beginning of the 1990s there was a set of national standards for human gene therapy proposals, and the proposals were subject to public review by an advisory committee and subcommittee at the NIH. In contrast, the technology of genetic testing and screening was regulated in part by the Food and Drug Administration, which must approve commercially available test kits as medical devices, and in part by a series of divergent statutes and regulations on the state level. There was, however, no public oversight body especially appointed for genetic testing and screening, and the elements of a coherent national strategy for this important technology were only beginning to emerge.

In the early 1980s, it might have seemed that human gene therapy was the technology requiring more urgent public attention. In June 1980, the general secretaries of the three largest religious bodies in the United States sent a letter to then-President Carter urging a detailed study of ethical issues in genetic engineering.[4] In July of the same year, Dr. Martin Cline of University of California at Los Angeles had, without prior authorization from his local Human Subjects Protection Committee, performed gene therapy on two patients, one in Israel and one in Italy.[5] Further, the issue of germline genetic engineering had received a significant amount of public attention, especially through the writings of Jeremy Rifkin.[6] Two results of the public concern about human gene therapy were a 1982 congressional hearing on Human Genetic Engineering and a 1982 comprehensive report, *Splicing Life,* by the President's Commission on Bioethics.

The congressional hearing, the *Splicing Life* report, and the NIH Recombinant DNA Advisory Committee set the agenda for public policies regarding human gene therapy in the decade of the 1980s. In a prescient concluding chapter entitled "Protecting the Future," the *Splicing Life* report considered

> . . . the means through which the issues generated by genetic engineering can continue to receive appropriate attention. They will be of concern to the people of this country—and of the entire globe—for the foreseeable future; indeed, the results of research and development in gene splicing will be one of the major determinants of the shape of that future.[7]

The Commission went on to state several observations based on its detailed study of recombinant DNA research:

[1] The issues raised by the projected human uses of gene splicing, which heretofore have not received attention, are at least as complex and important as those addressed by RAC thus far.

[2] These issues would benefit from an evaluation process that is continuing rather than sporadic, to allow a review body to develop coherent standards and orderly procedures, while making provisions for unexpected developments in gene splicing and other changes in the world at large.

[3] It would be desirable to develop means for such an evaluation process now, not because of any threat of iminent harm but because the issues are better addressed in anticipation rather than in the wake of a possible untoward or unforeseen outcome.

[4] The issues are so wide-ranging as to require a process that is broad-based rather than primarily expert, since the issues cannot be resolved on technical grounds alone and since many of the most knowledgeable scientists are deeply involved in the field as researchers or even as entrepreneurs.[8]

In response to the *Splicing Life* report, the NIH RAC created a national public review mechanism for human gene therapy proposals in 1983 and 1984. Some of the major events in the work of the RAC and its subcommittee can be summarized as follows:

1983: The RAC formed a working group to study the *Splicing Life* report; in response to the working group report, RAC members expressed their willingness, in principle, to review human gene therapy proposals.

1984–1985: A subcommittee to the RAC was established to provide initial review of human gene therapy proposals; the subcommittee drew up a series of guidelines for researchers called "Points to Consider."[9]

1987: The Human Gene Therapy Subcommittee reviewed a model protocol, "Human Gene Therapy: Preclinical Data Document," submitted by W. French Anderson and his colleagues.

1988–1989: The first human gene-marking (or gene-transfer) protocol, submitted by Steven A. Rosenberg and his colleagues, was reviewed and recommended to the NIH Director for his approval by the subcommittee and the RAC.

1990–1991: The first two human gene therapy protcols, submitted by R. Michael Blaese and W. French Anderson and by Steven A. Rosenberg, were reviewed and recommended for approval by the subcommittee and the RAC; the former protocol proposed to treat children with severe combined immune deficiency caused by adenosine deaminase deficiency; the latter protocol aimed to treat malignant melanoma patients by inserting the gene that produces tumor necrosis factor into some of the patients' white blood cells.[10]

Public policies for the technology of genetic testing and screening have evolved in a rather different way. In fact, it could be argued that the decade of the 1980s saw opposite movements in the gene therapy and the genetic testing and screening arenas. For somatic cell human gene therapy, at least, a kind of ethical closure had been reached by the end of the decade. In contrast, the public policy questions surrounding genetic testing and screening seemed more amenable to resolution at the beginning of the decade than at its end.

When the President's Commission published its detailed report on genetic counseling and screening in 1983, the policy questions surounding this technology seemed to be thoroughly understood and well under control. A crisis had occurred in the early 1970s around ill-advised programs of testing and screening for sickle cell trait and sicke cell disease.[11] However, position papers by a Hastings Center research group[12] and an influential report by a committee of the National Research Council[13] helped to reverse coercive and discriminatory policies. By the time of the Commission report, a professional and indeed national consensus seemed to have been reached on the major issues surrounding neonatal screening, prenatal diagnosis, and carrier testing. The disclosure of relevant genetic information to third parties without consent and prenatal diagnosis for sex selection remained controversial, and the potential development of a test for cystic fibrosis was a small cloud on the future horizon, but ethical closure had been reached on most issues within the testing and screening sphere.

By the early 1990s, this apparent closure threatened to come undone. In an insightful, widely cited essay, Alexander Capron argued that the "threat" of modern genetics was not posed by human genetic engineering proposals, or by the multibillion-dollar Human Genome Project; rather he stated, genetic screening is both the most probable and the most imminent source of harm to human beings.[14] Capron noted, in particular, the potentially adverse effects of genetic screening on the individual's insurability (especially for health care) and employability. He urged that several fundamental issues be elevated "to a level of much greater visibility"—among them our notions of normality and abnormality, our conceptions of the proper social roles for health insurance and employment, and our attempt to develop a public ethic of responsible parenthood.[15]

In contrast to the comprehensive and searching analysis called for by Capron, much of the ethical debate about genetic testing and screening since 1983 has been distinctly ad hoc in character. One thinks, for example, of the multi-year debate about the appropriate uses of the available test for Huntington disease (HD). On the heels of the HD debate, the discovery of the locus for the cystic fibrosis (CF) gene led to urgent consultations and a national consensus conference on the appropriate use of the not-yet-perfected CF test. Months later, a special consultation was held on the p53 gene and its potential role as a marker for increased susceptibility to certain cancers. Meanwhile, the Office of Technology Assessment has produced important studies on the forensic uses of DNA testing and on genetic monitoring and screening in the workplace.[16] Further, the National Center for Human Genome Research and the Human Genome Program in the Department of Energy have made approximately twenty-five awards to various academic and other institutions to support studies of, and educational programs related to, the ethical, legal, and social implications of the Human Genome Project. Most of the awards to date are either directly or indirectly concerned with genetic testing and screening.

Yet the task of achieving an overview of the multiple professional initiatives currently in progress and of synthesizing the results of multiple research projects will be formidable. The task is especially challenging when one considers that the fifty states and the District of Columbia play the predominant roles in both employment law and insurance regulation. A further factor that makes the study of genetic testing and screening quite complex is the involvement of numerous biotechnology and pharmaceutical companies. This commercial dimension of genetic testing and screening stands in sharp contrast to the early development of human gene therapy, which has been predominantly university-based.

Several initiatives in the direction of synthesis have already been undertaken. Neil Holtzman's important book, *Proceed with Caution,* provides a helpful guide to some of the important long-term issues surrounding genetic testing and screening, as they appeared in 1988.[17] The relevant staff officers at the National Center for Human Genome Research and the Department of Energy have achieved impressive personal syntheses on the normative issues, assisted by a distinguished NIH-DOE Working Group on Ethical, Legal, and Social Issues. Initial surveys of the major normative question have appeared in recent publications of the cooperating agencies.[18]

My intention in the preceding remarks and the next two parts of this chapter is not to minimize the importance of what has already been accomplished. It is, rather, to emphasize the magnitude of the task that lies before us and to argue for a national commitment of resources proportional to that task.

Major Goals for a National Advisory Committee

Classically, public advisory committees or commissions have had at least five major goals or performed five major roles.[19] The first is to "lend added dignity, authority, or apparent impartiality to official action to official action and thus . . . to legitimize both the action itself and the regime that established the commission."[20] One political scientist, David Flitner, has called this function of commissions "symbolic reassurance."[21] Second, commissions can perform an important bureaucratic role if they are "used as a tool for surmounting the pathologies of organizational complexity."[22] Third, if their members hold various viewpoints or espouse competing interests, commissions can perform a representational function. Fourth, if commissions perform their research roles well, they can develop the factual bases for informed public policies. Finally, through their reports and recommendations advisory commissions can educate the public and, if necessary, build public support for policy changes.[23]

Advisory committees or commissions with a designated life span—for example, three to five years—may be especially well suited to achieving goals like those outlined above. Such temporary bodies have at least the potential to approximate the academic ideals of comprehensiveness and impartiality. In his study of presidential advisory commissions from Truman to Nixon, Thomas Wolanin argued that temporary ad hoc commissions

> are uniquely capable of analyzing problems because they are temporary systems; they can recruit well-qualified members and staff; they have unusually good access to expertise and to data; and they serve as an integrative framework for an interdisciplinary and multi-interest consideration of problems. Commissions are also particularly capable of persuading others to accept as authoritative their findings and recommendations because they can command a wide audience for their reports; they can call upon the prestige of their members and of the [appointing authority] to increase the likelihood of a favorable hearing; they have a decision-making process that conforms to the public's ideal of how decisions should be made; and they enjoy the benefits of being both inside and outside the government.[24]

What, then, would be the specific goals for a National Advisory Committee on Genetic Testing and Screening? A first important goal for such a committee would be to monitor developments in biomedical research that could have important implications for genetic testing. The most important discoveries are usually published in major journals and widely reported by the press and the media. However, scientist-members of such a committee would be attuned to the international research network and could in many cases alert the committee to forthcoming publications. Further, the committee could systematically track new findings and classify them under general and specific headings like the following:

General Category	Specific Condition
Somatic illness	Cystic fibrosis
Mental illness	Schizophrenia
Vulnerability	Tendency to develop coronary artery disease

A second major goal of a national advisory committee would be less technical and more normative, namely, to deliberate about the ethical, legal, and social issues currently raised, or likely to be raised in the future, by new knowledge about human genetic diseases or susceptibilities. As noted in the *Splicing Life* report, this kind of deliberation will be more useful if it is anticipatory rather than a response to an immediate crisis.

A third goal of a national advisory committee would almost surely be public and professional education. Press and media coverage of the committee's meetings would, by itself, promote general knowledge of new developments and their social implications. The committee could also produce reports, brochures, and bibliographies targeted toward the general public, on the one hand, and health professionals, on the other.

Only a short step removed from its educational goal would be the goal of providing advice to professionals and policy makers. Such advice could be presented either in terms of multiple options (in the manner of the Office of Technology Assessment) or as a recommendation of what is considered to be the best policy or practice in a given set of circumstances.

A goal that, in my view, should not be adopted by such a national advisory committee is direct involvement in the process of legislation or regulation. The basis for this preference is the judgment that it is useful to maintain a division of labor between more academically oriented and relatively objective bodies like advisory committees, on the one hand, and the relatively more political work of legislating and regulating, on the other. At the same time, however, well-focused committees can stimulate the standard-setting work of professional societies or the National Conference of Commissioners on Uniform State Laws and can provide useful information to policy makers at the federal and state levels.

The Structure and Locus of a National Advisory Committee

There are several desirable general characteristics of a National Advisory Committee on Genetic Testing and Screening on which all thoughtful people could readily agree. It should be an interdisciplinary body comprised of men and women of diverse backgrounds and points of view. It should conduct its work entirely in public. It should be supported by a knowledgeable and committed staff. And the reports and recommendations of the committee should reflect careful, global research and should be clearly and coherently argued.[25]

More detailed specification of the structure and locus of a national advisory committee invites controversy. Yet, if the discussion of this important initiative is to advance, theses will need to be put forward and debated. It is in a spirit of exploration that I put forward the following concrete proposals.

One of the most fundamental questions surrounding the proposed committee is whether it should be a government-appointed or private-sector body. The principal arguments in favor of a government-appointed body are that it acquires a certain legitimacy and that its findings may therefore be regarded as authoritative. The principal arguments against a government-appointed committee are that it could be politicized or coopted or fall prey to both deficiencies simultaneously.

In its *Splicing Life* report, the President's Commission opted without justification for some kind of government-appointed body as the appropriate oversight mechanism for the entire field of recombinant DNA research. It then discussed the relative merits and liabiilties of a remodeled RAC, a genetic engineering commission, and a general federal biomedical ethics commission in performing the proposed responsibility. However, since 1982 there have been at least three instances in the United States and the United Kingdom in which private-sector initiatives have had a major impact in controversial areas of biomedical research or practice. These are the work of the Interim Licensing Authority for in vitro fertilization research and practice in the United Kingdom,[26] the *Confronting AIDS* report of the Institute of Medicine and National Academy of Sciences,[27] and the report of the American Fertility Society's Ethics Committee on the new reproductive technologies.[28] Simultaneously, some government-appointed bodies in the United States have foundered on seemingly intractable ethical and political disputes—most notably the congressional Biomedical Ethics Advisory Board and the NIH panel on human fetal tissue transplantation research. Thus, I conclude that both public- and private-sector approaches should be explored.

In the public sector, there would be two primary options for creating a National Advisory Committee on Genetic Testing and Screening. The first would be to modify the role of an already-existing group; the second would be to create a group de novo. The two existing groups that are working in the closest proximity to genetic testing are the NIH-DOE Working Group on Ethical, Legal, and Social Implications of Human Genome Research and the NIH RAC and its Human Gene Therapy Subcommittee. As the above chronology illustrates, the RAC has already demonstrated sufficient flexibility to make a major transition from the laboratory to the clinic. The NIH-DOE Working Group has a less extensive track record than the RAC or its subcommittee, and its deliberations to date have been less well publicized than those of the RAC. Yet the Working Group has already devoted considerable attention to normative questions surrounding genetic testing and screening. One possible strategy would be to ask the Department of Health and Human Services and/or the Department of Energy to devote a small fraction of their considerable resources to the support of a new national advisory body that builds on the strengths of one or both of these existing groups.

Alternatively, a new advisory committee could be created within the federal government. As we have seen, a Genetic Engineering Commission with a considerably broader mandate was suggested as an option by the President's Commission on Bioethics. Closer to the scope and function envisioned in this essay would be a President's Commission on the Human Applications of Genetic Engineering; then-Congressman Albert Gore proposed the creation of such a body in April 1983.[29] One possible advantage of such a new committee is that it could start afresh, with a clear mandate presumably established by the Congress and a clear public perception of independence from executive branch agencies. Possible disadvantages of this model could be prolonged debate about the appointment of committee members and the intrusion of abortion politics into the appointment and agenda of the committee.

It would be more difficult for a private-sector committee to achieve either the visibility or the credibility of a publicly appointed committee, yet there are precedents for this approach to controversial topics. Perhaps the most interesting model is the Interim Licensing Authority (ILA) in the United Kingdom. When Parliament hesitated to follow up on the recommendations of the Warnock Committee report of 1984, the Royal College of Obstetricians and Gynaecologists, with funding from the government's Medical Research Council, established the ILA to oversee both laboratory research with human preimplantation embryos and the clinical practice of in vitro fertilization. The ILA had no real enforcement authority but was able to use both moral suasion and public reporting to achieve rather uniform standards of practice throughout the United Kingdom.

A similar panel has been launched by the American Fertility Society, the American College of Obstetricians and Gynecologists, and several other professional societies in the United States. These groups, convinced that the federal government will not soon reestablish the DHHS Ethics Advisory Board, established a National Advisory Board on Ethics in Reproduction (NABER), which will provide eternal review of research protocols involving, for example, preimplantation embryos. In the case of genetic testing and screening, the

American Society of Human Genetics, the American Academy of Pediatrics, and the American College of Obstetricians and Gynecologists would be three highly interested professional groups. Disease-oriented voluntary associations like the National Organization for Rare Disorders and the Cystic Fibrosis Foundation would also want to be actively involved. The Howard Hughes Medical Institute, the Department of Energy, and the National Center for Human Genome Research could be approached for partial funding of the committee.

Conclusion

There is a need for a National Advisory Committee on Genetic Testing and Screening. This chapter has suggested several important goals for such a body, and outlined three possible models, two in the public sector and one in the private sector. Where, how, and by whom a national advisory committee is created are secondary and instrumental questions. What is most important is that a competent committee be established, that it do its work well, and that it begins its work soon.

ACKNOWLEDGMENTS

The author's research for this chapter was supported in part by the Biomedical Research Support Grant program of the National Institutes of Health (S07RR07136-18). The author is also indebted to his colleagues in the National Reference Center for Bioethic Literature and the Bioethics Information Retrieval Project for their collection and indexing of materials used in the preparation of the chapter.

NOTES

1. B. Davis, "Prospects of Genetic Intervention in Man," *Science* 170(964):1279–83, 1970; M. P. Hamilton, ed., *The New Genetics and the Future of Man,* Eerdmans Publishing Co., Grand Rapids, MI, 1972.
2. President's Commission for the Study of Ethical Problems in Medicine and Biomedical and Behavioral Research, *Screening and Counseling for Genetic Conditions: The Ethical, Social, and Legal Implications of Genetic Screening, Counseling, and Education Programs,* U.S. Government Printing Office, Washington, D.C., 1983, pp. 11–31.
3. President's Commission for the Study of Ethical Problems in Medicine and Biomedical and Behavioral Research, *Splicing Life: The Social and Ethical Issues of*

Genetic Engineering with Human Beings, U.S. Government Printing Office, Washington, D.C., 1982.

4. Ibid., pp. 95–6.
5. P. Jacobs "Pioneer Genetic Implants Revealed," *Los Angeles Times,* October 8, 1980, pp. 1, 26, 27.
6. T. Howard and J. Rifkin, *Who Should Play God?* Dell Publishing Co., New York, 1977.
7. See President's Commission, 1982, p. 81.
8. Ibid., pp. 81–2.
9. U.S. National Institutes of Health, "Points to Consider in the Design and Submission of Human Somatic-Cell Gene Therapy Protocols," *Federal Register* 50:33463–7, (August 1985).
10. L. Walters, "Human Gene Therapy: Ethics and Public Policy," *Human Gene Therapy* 2(2): 115–22, 1991.
11. P. Reilly, *Genetics, Law, and Social Policy,* Harvard University Press, Cambridge, MA, 1977, pp. 62–86.
12. Institute of Society, Ethics, and the Life Sciences, Research Group on Ethical, Social, and Legal Issues in Genetic Counseling and Genetic Engineering, "Ethical and Social Issues in Screening for Genetic Disease," *New England Journal of Medicine* 286:1129–32, 1972; T. M. Powledge and J. C. Fletcher, "Guidelines for the Ethical, Social, and Legal Issues in Prenatal Diagnosis," *New England Journal of Medicine* 300:168–72, 1979.
13. National Research Council, Committee for the Study of Inborn Errors of Metabolism, *Genetic Screening: Programs, Principles, and Research,* National Academy of Sciences, Washington, D.C., 1975.
14. A. M. Capron, "Which Ills to Bear? Reevaluating the 'Threat' of Modern Genetics," *Emory Law Journal* 39:665–6, 1990.
15. Ibid., pp. 693–6.
16. U.S. Congress, Office of Technology Assessment, *Genetic Witness: Forensic Uses of DNA Tests,* OTA, Washington, D.C., 1990; U.S. Congress, Office of Technology Assessment, *Genetic Monitoring and Screening in the Workplace,* OTA, Washington, D.C., 1990.
17. N. A. Holtzman, *Proceed with Caution: Predicting Genetic Risks in the Recombinant DNA Era,* Johns Hopkins University Press, Baltimore, 1989.
18. U.S. Department of Health and Human Services and U.S. Department of Energy, *Understanding Our Genetic Inheritance: The U.S. Human Genome Project: The First Five Years, FY 1991–1995,* Departments of Health and Human Services and Energy, Washington, D.C., 1990.
19. LeRoy Walters, "Commissions and Bioethics," *Journal of Medicine and Philosophy* 14:363–8, 1989.
20. H. C. Mansfield, "Commissions, Government," in *International Encyclopedia of the Social Sciences,* Vol. 3, Macmillan-Free Press, New York, pp. 12–18.
21. D. Flitner, Jr., *The Politics of Presidential Commissions: A Public Policy Perspective,* Transnational Publications, Dobbs Ferry, NY, 1986.
22. Ibid., p. 23.
23. Walters, 1989, pp. 364–5.
24. T. R. Wolanin, *Presidential Advisory Commissions: Truman to Nixon,* University of Wisconsin Press, Madison, 1975.

25. B. A. Brody, "The President's Commission: The Need to Be More Philosophical," *Journal of Medicine and Philosophy* 14:369–83, 1989; and Walters, 1989.
26. Interim Licensing Authority for Human In Vitro Fertilisation and Embryology, *A Report on Research Licensed by the Interim Licensing Authority (ILA) for Human In Vitro Fertilisation and Embryology, 1985–1989,* ILA Secretariat, c/o Medical Research Council, London, 1989.
27. Institute of Medicine and National Academy of Sciences, Committee on a National Strategy for AIDS, *Confronting AIDS: Directions for Public Health, Health Care, and Research,* National Academy Press, Washington, D.C., 1986.
28. American Fertility Society, Ethics Committee, "Ethical Considerations of the New Reproductive Technologies," *Fertility and Sterility* 53(Suppl. 2):1S–109S, 1990.
29. U.S. Congress, "A Bill to Establish the President's Commission on the Human Applications of Genetic Engineering," H.R. 2788, 98th Congress, 1st Session, 1983.

VI

Conclusion

17

Social Policy Research Priorities for the Human Genome Project

George J. Annas and Sherman Elias

It is fair to conclude that the legal, social, and ethical issues highlighted in this book have received almost no attention from the scientific community involved in the Human Genome Project. The announcement for the third annual national conference, Human Genome III (held in October 1991), for example, touted the project as "the largest biological project ever contemplated on the status and future of research on the human genome." Six sessions were held over the three days of the meeting. All but one were devoted exclusively to science; the only exception was on "politics." Changing the perception that there are only scientific (and some political) issues at stake in the Human Genome Project will not be easy—and may not be possible. In this regard the Human Genome Project's social policy research program is certainly necessary, but it should not be seen as sufficient.

Previous attempts to develop social policy in anticipation of scientific advances have not been successful. From nuclear power to space exploration, and from organ transplantation to the AIDS epidemic, law and social policy have almost always been made in reaction to scientific discoveries and technological developments. The Human Genome Project's early approach of earmarking a portion of its research budget for the study of "ethical, social, and legal issues" is unprecedented and could provide a model for other federally funded scientific research. A systematic study of social policy issues related to technology development provides an extraordinary opportunity to test the hypothesis that responsible social policy can be developed prior to or simultaneously with scientific research and technological advance.

The Human Genome Project, however, is especially commendable for this social policy development experiment. The Project promises great rewards in both basic scientific knowledge and clinical medicine. Our century's experience of genetic and racial discrimination means that new social policy approaches to genetics will be necessary to safeguard human rights. Even though genetics has a unique history, the social policy approach should be generalizable because there are no ethical, legal, or social policy issues that are unique to the Project.

The Project's social policy program also permits exploration of a wide range of unresolved social policy issues in both scientific research and clinical medicine which are relevant to the Human Genome Project. These issues are well summarized in this volume. But the purpose of bringing these scholars together was more than just to produce an up-to-date, state-of-the-art document. If the social policy experiment is to have a fair chance at success, an early consensus must be developed concerning the social policy issues that warrant top priority for funding and research. Without priority setting, this program will quickly become simply an "open" funding source for everyone's favorite project that could not obtain financing elsewhere, but could somehow relate itself to the Human Genome Project.

At a priority-setting workshop in early 1991, the contributors to this volume all agreed that the Human Genome Project has been the subject of misleading and counterproductive hyperbole. Damage control in the form of more realistic descriptions of the project and its goals are in order. There are many "Holy Grails" in science today, and the search for the meaning of life is as alive in astronomy as in molecular biology. Nor can the Human Genome Project solve all of society's problems. For example, issues of genetic discrimination in health insurance are a central concern. However, these can be effectively dealt with only by providing access to health care to all Americans, most likely through a system of national insurance. Designing a universal health insurance system is one of the country's top domestic issues, but is not one to which the Project has much to add directly. Likewise, military applications of the fruits of the Human Genome Project may ultimately prove the most socially dangerous. Nonetheless, it is unlikely that social policy researchers funded by either NIH or the Department of Energy (DOE) would be able to add much to the military application discussion. A ban on biologically based weapons and defense systems seems most prudent (based on what we already know), but enforcement is problematic. Education of both the public and health care professionals about the social policy issues and options seems essential. Even though the Human Genome Project itself may be the proper agency to fund some educational programs, it will undoubtedly be accused of being self-serving in this endeavor.

The remaining issues were reviewed by the contributors for their importance to developing social policy. It was concluded that the most pressing issues involve the medical applications of the products of the Human Genome Project, which are likely to greatly increase the number of genetic screening tests available. Cystic fibrosis (CF) screening, and the unresolved problems it poses, serves a useful model for exposition, even though the discovery of the primary genetic site responsible for this condition was made independent of the Human Genome Project. Thus it seems appropriate that the Human Genome Project has decided to help fund pilot studies on CF screening to develop public policy in this area. The contributors to this book and the other workshop participants (Glantz, Grodin, Lander, and Wexler), ultimately agreed that the following basic issues should be top research priorities, in the following order:

1. When and how should new genetic tests be introduced into medical practice?

This was the clear first-priority preference of the group. It is a predictable preference for the physician participants, who worry about keeping up with new technology, maintaining ethical standards, and avoiding medical malpractice suits. This basic issue involves a number of subissues. The first is how the medical standard of care should be set. Historically, physicians have set their own professional practice standards, and courts have relied on medical custom to define "reasonable prudence," and thus the standard of care. When medical technology changes as quickly as it currently does, and when medical malpractice cases may take five to eight years to resolve, litigation is an inefficient and ineffective way to determine the standard of medical care.

Most genetics litigation has also been limited to the area of prenatal screening (for such conditions as Down syndrome and Tay-Sachs disease). Wrongful birth (parents suing for negligent failure to provide adequate prenatal screening and counseling) and wrongful life (children suing on the basis that they would have been better off not having been born) suits will continue to be a source of standard setting in the absence of effective professional standard setting with public input. The abortion controversy will also remain central to the introduction of any new prenatal screening test, at least until method are developed to treat genetic abnormalities that now are the basis for decisions to terminate a pregnancy. The most socially problematic screening tests, however, will be those that detect presymptomatic conditions or conditions that will only develop in a subset of those who carry particular genes. Unresolved issues include when, if ever, such screening should become routine or standard for neonates, school-age children, or adults, who should determine this, and on what basis.

A second subissue is what information patient/clients should be given regarding any particular screening test, how the information should be presented, when it should be presented, and who should present it. Experience with human immunodeficiency virus (HIV) testing, for example, suggests that counseling must occur even before the test is performed, that the test should be entirely voluntary, and that counseling must also be an integral part of conveying the results of the test. On the other hand, HIV screening may not provide the proper model. Since it is likely that hundreds of screening tests will eventually be developed, it will literally be impossible to do meaningful counseling about each and every one of them. Accordingly, we have suggested that there will be a need to develop "generic genetic counseling" to provide sufficient information to permit individuals to give informed consent to screening, but not so much information as to amount to misinformation that would serve only to make the process misleading or meaningless.

It has been popular at the outset of the availability of CF screening to "inform" patients of the existence of the test, even if the physician or counselor does not recommend the test. Is this ever proper? If the model is consumer based, then perhaps we should conclude that "the customer is always right," and let individuals decide. On the other hand, if the proper model is physician-patient, might it not properly be considered a breach of fiduciary duty to inform a patient of the existence of a screening test that the physician is not prepared to recommend? Both models exist in the literature, but they lead to radically different conclusions about counseling, and we will have to choose one or the other, or invent a third.

2. How can the confidentiality and privacy of an individual's genetic information be preserved?

Because genetic information can be used to stigmatize individuals and enable others to make discriminatory decisions about them, and because much genetic information is potentially self-defining and sometimes embarrassing, individuals have a strong interest in keeping it secret. Maintaining the security of genetic records will be essential if genetic screening is to achieve public acceptance. Current procedures for maintainnig the confidentiality of medical records are woefully inadequate, and new strategies will have to be designed.

Two specific, unique record-keeping problems must also be systematically addressed. The first relates to informing other family members of genetic traits that they may carry. When a physician's patient threatens to kill a third person, the physician has a duty to protect that permits disclosure. But the circumstances under which a physician may disclose genetic information that may affect another family member or future children have never been adequately detailed.

The second issue involves so-called DNA or "gene banks" in which samples of a person's blood or other tissue are maintained for future use of comparison. This record is similar to a medical record in that it is usually maintained at a health care facility and was obtained with a health care purpose. Unlike a typical medical record, which can be seen as a partial past diary, the DNA record contains a partial "future diary." No one knows what it contains—since each new genetic discovery (or decoding) permits new information to be obtained from the DNA sample. Criminal justice officials insist that their DNA records are only for identification purposes (like fingerprints), but such limited use cannot be assured even with current technology. It is essential that the record-keeping rules for DNA banks be spelled out *before* individuals are asked to have their samples stored. These rules are likely to closely follow the recommendations made in the mid-1970s for computerized medical records, but should be more stringent. For example, because no one can know what the record is likely to contain in the future, a rule that gives individuals an absolute right to order their stored sample destroyed at any time seems reasonable. For a complete set of rules the editors recommend for gene banks, see pp. 9–10.

3. How can genetic discrimination by employers and insurance companies be prevented?

It is likely that both employers and insurance companies will want to use large-scale genetic screening of applicants when the technology becomes relatively inexpensive. Since nongroup insurance is based on weeding out bad risks, such as persons with preexisting conditions (for example, hypertension), screening out individuals with "bad genes" will be natural and acceptable under current law and underwriting standards. The challenge is to prevent the products of the Human Genome Project from becoming just another mechanism for discrimination. For employers who provide health insurance for their workers, the issue of genetic discrimination is substantially similar; that is, employers will want to exclude individuals who may cost them large sums of money in terms of future health bills. In addition to this, however, employers may also want to exclude workers who may die prematurely or develop early disease because of the cost of training, the need for experienced workers, or other job-related rationale. Accordingly, not only confidentiality and privacy, but also the rights of individuals to refuse such screening or to prevent it from being used against them, become central. Since genetic predispositions are not dealt with specifically in the Americans with Disabilities Act or elsewhere, this social policy issue remains unresolved. It is especially difficult because not all genetic predispositions will ever manifest themselves, and, even when they do, will do so to widely-varying degrees.

4. How might the Human Genome Project affect our concepts of "disease," "normalcy," and "humanness"?

Under what circumstances should a gene be considered evidence of a disease or abnormality, and what difference does it make if the genotype is not expressed or has only a statistical probability of ever being expressed? Many commentators have worried that inventorying our genomes for defects, mistakes, or abnormalities may make us feel more fragile and vulnerable; it may make us all think of ourselves as sick. What does it mean to have an "abnormal" genome? Related to concepts of health and disease are concepts of reductionism and determinism. We know that we are more than the sum of our genes, but a concentration on our genetic composition may make us think of ourselves primarily as a composite of genes and lead us to marginalize the contribution of environment. The 47,XYY experience has also taught us that a genetic explanation for behavior can be both powerful and misleading.

Related to all these notions is the overriding issue of eugenics and what society should do, if anything, to either improve the genetic composition of its members or decrease the genetic burden. This issue, in turn, is related to society's general problems with racism, sexism, and poverty. As with all expensive, new medical technology, the fruits of the Human Genome Project are likely to go primarily to the wealthy; its stigma potential is likely to be used primarily against the poor.

Other Issues

Other issues were also thought to warrant serious study, but it was more difficult to achieve a consensus on their priority. Commercialization, especially patenting the Genome Project's products, and government-industry cooperation to establish and maintain U.S. primacy in the biotechnology field, ranked high. In retrospect, patenting issues should probably have been ranked first. They have become *the* most controversial of all social policy issues in the Human Genome Project. Patenting parts of the human genome not only threatens to end current international cooperation in genetics research, but has also directly led to the April 1992 resignation of James Watson as head of NIH's Genome Center in a policy dispute with NIH Director Bernadine Healy over whether NIH should now seek to patent cDNA—a policy Healy supports over Watson's strenuous objections. Neither somatic cell enhancement or germline gene therapy issues ranked high because much work has already been done on these issues, and regulatory mechanisms are in place to protect the human subjects of research trials. Nonetheless, the proper role of federal regulatory agencies in setting standards for such techniques and determining the conditions under which they should be made generally available as part of a minimum health care package have not received sufficient attention.

It is of interest that the workshop's three top priorities for the social policy studies portion of the Human Genome Project are substantially similar to the three priority areas identified by the Project's own Working Group for Ethical, Legal, and Social Implications (ELSI) in late 1990. Three of the members of that group (King, Murray, and Wexler) were participants in our workshop and, of course, brought their prior experience with them. On the other hand, they were invited long before they were named to ELSI. Moreover, they were the only ones conversant with their earlier list, and this list was not shared with our participants until very late in the workshop. In addition, the workshop approached the priority-setting issue from an entirely different perspective, based on the level of concern (individual/societal/species) rather than particular issues. Thus we think it is fair to conclude that at the outset of the HGP, there is a professional consensus on research priorities in the social policy area of the Human Genome Project. Nonetheless, no priority list is eternal, and a similarly constituted outside group should revisit this issue at least once every two years, both to evaluate the work that has been done and to determine the continued validity of the research priorities.

Perhaps the most important social policy issue of all—should the Human Genome Project proceed at this time?—received no priority rating. This is unremarkable. The Project itself is not the appropriate funder for any research designed to give an "independent" or "objective" assessment of its own priority in scientific research. This type of assessment must await independent funding of neutral researchers. In this regard, we found workshop participant Eric Lander's response to the question "Will the Human Genome Project distort research for molecular biology?" both instructive and accurate. His response: "It is much more likely to distort research in bioethics."

We have already observed the ability of a large funding source, the largest in the history of U.S. bioethics, to attract proposals from almost every serious scholar and researcher in the field. Just as the focus on genes may, at least in the short run, persuade us to see all of society's problems as gene-related, so could bioethicists persuade themselves and others that the only important social policy questions in bioethics involve genetics. A prioritized research agenda should help us focus our energies on real problems affecting real people, without abandoning our concern for broader social policy issues and fundamental philosophical questions. Scientists and policy makers alike have a large stake in the success of the Human Genome Project's social policy development experiment. Social policy development is an integral part of modern science and, done properly, could act as a key promoter for scientific development. But it will take more than good intentions, and even more than adequate funding: both the scientists and the public must get involved in open and intense discussions of the issues outlined in this volume if human rights and human dignity are to survive the genetic revolution. We hope this book plays a constructive role in helping to promote and focus that discussion.

Glossary

Allele   an alternative form of a gene, occupying a specific site (locus) on a chromosome.

Amino Acid   one of the twenty different molecules that combine to form proteins. The specific sequence of amino acids in a protein is determined in the genetic code of DNA.

Amniocentesis   a prenatal diagnostic procedure that involves withdrawing a small sample of amniotic fluid from the amniotic sac surrounding the fetus. Although amniocentesis has traditionally been performed around sixteen weeks of pregnancy, obstetricians have recently begun performing the procedure earlier in pregnancy.

Base   one of the five nitrogen molecules that make up the informational content of DNA and RNA. DNA contains four bases that pair across the chains of the double helix: adenine always pairs with thymine, and cytosine always pairs with guanine. RNA is single-stranded and contains adenine, guanine, cytosine, and uracil.

Base Pair   pairs of complementary bases forming the DNA double helix. The double-stranded molecule has a structure like a twisted ladder, in which each rung consists of a nucleotide base pair held together by weak hydrogen bonds. The base pairs consist of adenine (A), which must always pair with thymine (T), and guanine (G), which must always pair with cytosine (C). Base pairs are the units used to measure the length of DNA.

Centimorgan   a unit of genetic distance that measures recombination frequency. One centimorgan is equal to a 1 percent chance that a genetic locus will be separated from a genetic marker due to recombination in a single generation. In humans, on average one centimorgan is equivalent to a physical distance of one million base pairs.

Chorionic Villus Sampling   a prenatal diagnostic procedure, usually performed around ten to twelve weeks of pregnancy, whereby a small sample of the placenta (villi) is obtained by passing either a catheter (small tube) through the cervix or a needle through the abdomen. Both procedures are performed under ultrasound guidence.

Chromosome   subcellular structures within the nucleus of the cell, which are composed of DNA and proteins. In humans there are a total of twenty-three pairs of chromosomes, which can be visualized through the microscope when cells are specially prepared during certain stages of cell division. Genes are arranged in linear order along the chromosomes.

Clone   (1) verb: creating a group of genetically identical cells descended from a common ancestor cell; (2) noun: a group of genetically identical cells, all derived through cell division (mitosis) from a single ancestral cell.

Codon   a set of three bases (triplet) in a DNA or RNA molecule that codes for one of the twenty amino acids.

Complementary DNA (cDNA)   DNA synthesized from a messenger RNA template, using the bacterial enzyme reverse transcriptase.

Contig   a group of clones that represent contiguous (overlapping) regions of a genome.

Cosmid   a cloning vector consisting of a virus particle into which a DNA strand is packaged. A segment of DNA about 40,000 base pairs long can be inserted into these vectors.

Cytogenetics   the scientific study of the relationship between chromosomal structure and normal and pathological conditions.

DNA   (deoxyribonucleic acid) the molecule that encodes genetic information. DNA is a double-stranded molecule held together by weak hydrogen bonds between base pairs of nucleotides. DNA has a structure resembling a twisted double helix.

DNA Probe   a small segment of single-stranded DNA labeled (usually with a radioactive substance) so that it is detectable by means of hybridization. Probes can pinpoint complementary DNA sequences.

Dominant   a genetic trait that is expressed even though it is present in only a single copy (that is, in only one allele). Thus, the trait is expressed even when it is inherited from only one parent.

Electrophoresis   separation of molecules in solution (such as DNA fragments or proteins) on the basis of physical properties such as size and electrical charge. An electrical current is passed through a medium containing the mixture of molecules and each molecule travels through the medium at a different rate.

Enzyme   a protein that catalyzes (speeds up) a specific chemical reaction without altering its direction or nature.

Eukaryote   a cell or an organism characterized by having its chromosomes contained within a well-defined nuclear membrane. Eukaryotes include all organisms except viruses, bacteria, and blue-green algae, which are termed prokaryotes.

Exon   the protein-coding DNA sequences of a gene.

Gamete   a sperm or ovum.

Gene   a small section of DNA that contains information for the formation of a single-protein molecule. In some cases, a gene contains the information for making a molecule of transfer RNA or ribosomal RNA.

Gene Mapping   assigning the ordered relationships and distances between different genes or DNA segments on a chromosome.

Genome   all of the genetic material in the chromosomes of a particular organism or species. In humans, the genome consists of approximately three billion base pairs.

Genomics   a generic term for mapping and sequencing DNA.

Haploid   the chromosome number of a normal gamete (sperm or ovum), with only one member of each chromosome pair. In humans, the haploid number is twenty-three.

Heterozygous   possessing two different alleles at a specific locus on a pair of homologous chromosomes; having inherited different variants of a gene from each parent.

Homozygous   possessing a pair of identical alleles at a specific locus on a pair of homologous chromosomes; having inherited the same variants of a gene from each parent.

Human Genome Initiative   the collective name given to various projects directed toward (1) mapping and sequencing the human genome and genomes of other species, (2) developing of new computational methods for analyzing the genome (usually involving sophisticated computer technology), (3) development of new instruments and techniques for analyzing DNA.

Human Genome Project   the U.S. federally funded initiative to map and sequence the entire three billion base pairs of the human genome.

Hybridization   the process of joining two complementary strands of DNA, or DNA and RNA, to form a double-stranded molecule.

Introns   DNA sequences that interrupt protein-coding sequences of a gene. After information from the gene is transcribed into a new strand of messenger RNA, the introns are removed before the mature messenger RNA leaves the nucleus.

Kilobase   one thousand base pairs in a DNA sequence.

Library   as related to molecular genetics, a collection of clones of DNA segments.

Linkage   the proximity between two markers (for example, restriction fragment length polymorphism markers, genes) along a chromosome. The closer together genes are located, the less likely they are to separate during reproduction and, hence, the greater the probability they will be inherited together.

Locus   the precise location on a chromosome of a DNA segment or gene.

Lod Score   the logrhythm of the odds of linkage as opposed to nonlinkage of two genes. By convention, a lod score of +3 (1,000:1 odds) is taken as proof of linkage; a score of -2 (100:1 odds against) indicates no linkage.

Messenger RNA (mRNA)   the ribonucleac acid molecule that serves as a template to transmit the genetic information from the nucleus to the cytoplasm, where it directs the synthesis of a specific amino acid sequence of a protein.

Mutagen   a chemical or physical agent (for example, x-ray) that causes a permanent change in the DNA.

Mutation   a permanent change in the number, arrangement, or molecular sequence in DNA that can be inherited.

Nucleotides   a subunit of DNA or RNA consisting of a nitrogenous base, a sugar molecule (deoxyribose in DNA or ribose in RNA), and a phosphate molecule.

Phenotype   the observed result of interaction between genotype and environmental factors; the observable expression of a particular gene or genes.

Polymerase Chain Reaction   a technique for amplifying a short stretch of DNA. The method depends on the use of two flanking oligonucleotide DNA primers and repeated cycles of extension using the enzyme DNA polymerase.

Oligonucleotide   a short string of nucleotides.

Oncogene   a gene that plays an important role in the development of cancer.

Polymorphism   a difference in DNA sequence between individuals; an inherited genetic variation, such as the ABO blood groups.

Restriction Enzymes   enzymes purified from bacteria that cut double-stranded DNA at a specific nucleotide sequence site.

Restriction Fragment Length Polymorphism (RFLP)   a variation of DNA fragment sizes cut by restriction enzymes. Polymorphic sequences that are responsible for RFLPs are used as markers on genetic linkage maps.

Ribosomal RNA (rRNA)   a specific class of RNA found in the ribosomes of cells.

Ribosome   a subcellular structure composed of ribosomal RNA and proteins upon which proteins are created.

Somatic Cell   any cell in the body other than reproductive cells and their precursors.

Teratogen   a chemical or physical agent that produces birth defects.

Transcription   the synthesis of a single-stranded RNA molecule from a double-stranded DNA template; the first step in gene expression.

Transfer RNA (tRNA)   a specific type of RNA that brings amino acids into position on ribosomes to interact with messenger RNA for protein synthesis.

Translation   the process of synthesizing a polypeptide by converting the information in messenger RNA into protein. Translation follows transcription.

Yeast Artificial Chromosomes (YAC)   large segments of DNA (200,000 to 600,000 base pairs) from another species spliced into the DNA of certain types of yeast. YACs are used in mapping the genome of different species.

Index